Super Optical Biosensors

IOP Series in Sensors and Sensor Systems

The IOP Series in Sensors and Sensor Systems includes books on all aspects of the science and technology of sensors and sensor systems. Spanning fundamentals, fabrication, applications and processing, the series aims to provide a library for instrument and measurement scientists, engineers and technologists in universities and industry.

The series seeks (but is not restricted to) publications in the following topics:

- Advanced materials for sensing
- Biosensors
- Chemical sensors
- Industrial applications
- Internet of Things
- Lab-on-a-chip
- Localization and object tracking
- Manufacturing and packaging
- Mechanisms, modelling and simulations
- MEMS/NEMS
- Micro and nanosensors
- Non-destructive testing
- Optoelectronic and photonic sensors
- Optomechanical sensors
- Physical sensors
- Remote sensors
- Sensing for health, safety and security
- Sensing principles
- Sensing systems
- Sensor arrays
- Sensor devices
- Sensor networks
- Sensor technology and applications
- Signal processing and data analysis
- Smart sensors and monitoring
- Telemetry

Authors are encouraged to take advantage of electronic publication through the use of colour, animations, video and interactive elements to enhance the reader experience.

Super Optical Biosensors

Caide Xiao
Yunyang Medical College, China

IOP Publishing, Bristol, UK

Supplementary MathCAD programs and multimedia are available at http://iopscience.iop.org/book/978-0-7503-2384-0.

ISBN 978-0-7503-2384-0 (ebook)
ISBN 978-0-7503-2382-6 (print)
ISBN 978-0-7503-2383-3 (mobi)

DOI 10.1088/2053-2563/ab0823

Version: 20190501

IOP Expanding Physics
ISSN 2053-2563 (online)
ISSN 2054-7315 (print)

British Library Cataloguing-in-Publication Data: A catalogue record for this book is available from the British Library.

Published by IOP Publishing, wholly owned by The Institute of Physics, London

IOP Publishing, Temple Circus, Temple Way, Bristol, BS1 6HG, UK

US Office: IOP Publishing, Inc., 190 North Independence Mall West, Suite 601, Philadelphia, PA 19106, USA

This book is dedicated to my wife, Rongju Sun, and our son, Guangzhi Xiao.

Contents

Abstract

Surface plasmon resonance (SPR) biosensors, ellipsometry (EP) biosensors, resonant mirror (RM) biosensors, and interferometry biosensors, are optical instruments used to detect ligand–receptor interactions for biological research, clinical diagnoses and the pharmaceutical industry. They are not sensitive enough for samples with molecular weights less than a few kilo-Daltons. A hybridization of SPR and EP makes it possible to build a super biosensor SPREP which can be used to detect samples with small molecules (less than a few hundred Daltons), but the structure and operations of SPREP biosensors are very complex. In this book, the author presents two more super optical biosensors. The most sensitive super optical biosensor is named 'RMEPA', an abbreviation of three optical technique names: resonant mirroring, ellipsometry and attenuated total reflection (ATR). The structure of the RMEPA biosensor is so simple that even an incident angle scan system for a monochromatic light is not needed. By monitoring reflected light intensities at a fixed incident angle in total reflection, the phase angle difference between two polarized light waves can be extracted with an explicit formula. A sensitivity benchmark (γ) is introduced for comparisons of seven optical biosensors: SPR, EP, RM, porous silicon interferometry, SPREP, RMEP and RMEPA. The benchmark is a ratio of an optical biosensor's response to its sample's effective optical thickness. The RMEPA biosensor is found to be 30 times as sensitive as a commercial SPR biosensor, BIAcore 1000.

Preface

I began to study the surface plasmon resonance (SPR) biosensing technique in 1991 when I was a postgraduate student in Tsinghua University, Beijing, China. At that time, my laboratory lacked sufficient funds to buy expensive instruments. A null ellipsometer (EP) was used at that time to study Langmuir–Blodgett films and antigen–antibody interactions. Although EP was more sensitive than SPR, manual operations of the null ellipsometer were not quick enough to trace ligand–receptor interactions. Then I focused on SPR and made an automatic SPR biosensor for my PhD thesis. This book is a collection of my research over the last 28 years in many laboratories around the world. A new method to measure a phase angle difference of two linear polarized light waves in total reflection is presented in the book. It is a development of an extinct optical sensing technology resonant mirror (RM). In traditional ellipsometry, the symbol Δ is used for the phase angle difference, but in this book Φ is used for the same parameter because Δ is used as a symbol for any variable increment. By a hybridization of resonant mirror and ellipsometry, we can build a super optical biosensor named RMEP. A RMEP biosensor with two polarizers in fixed positions does not need a quarter-wave plate used in ellipsometry biosensors, and it can measure the phase angle Φ by monitoring the reflected light intensity of the system. Combining the attenuated total reflection (ATR) technique with RMEP, we can build a more powerful optical biosensor called RMEPA. The RMEPA biosensor is found to be 30 times more sensitive than a commercial SPR biosensor: BIAcore 1000. A benchmark (γ) for sensitivity and linearity comparisons of optical biosensors is proposed in the first chapter of the book. Since this book is written for readers with a physics background, there is a section in chapter 1 to introduce protein structures and ligand–receptor interactions. Protein such as antibodies and avidin are samples used in biosensors. In chapters 2–5, theories of SPR, ellipsometry, resonant mirror and porous silicon-based optical interferometric sensing techniques are introduced by solving Maxwell's equations in multilayered systems. Virtual optical biosensors were investigated by numeric simulations with Mathcad® programs to find benchmarks of those biosensors. These Mathcad programs are available for readers to download. In the last chapter, three home-made SPR biosensors demonstrate to readers that it is possible to create DIY optical biosensors. It has been a long journey to write the book. The idea to write a book about optical biosensors came to me in the year 2000 when I was a visiting scholar working in a research institute of the National Research Council Canada. At that time, I found it was necessary to set an absolute benchmark for sensitivity comparisons of different optical biosensors. It is just the same as the requirement for an absolute temperature scale to measure hot and cold. The idea of two virtual super optical biosensors RMEP and RMEPA came to me in February, 2018, and I am very happy to present them in section 5.3.

Caide Xiao
Calgary, Canada
December 16, 2018

Acknowledgments

To have my first English book to be published, I had much help. I wish to express sincere appreciation to Professor Senfang Sui. From his laboratory in Tsinghua University, I started my research career, and learnt how to write programs in Mathcad (version 3.1) for numerical simulations of my research projects. Mathcad is my virtual laboratory. After finishing my PhD study in Beijing, Dr Pier L San Biagio in Palermo, Italy offered me an opportunity to work in his laboratory of 'CNR-Istituto di Biofisica' as a visiting scholar from 1997 to 1998. This was very important for me to continue my research career. During that one-year time, I saw the beautiful island of Sicily, met many kind Italians, and tasted the best coffee and pizza on Earth. This book would not have been possible without help from the professional staff at Institute of Physics (IOP) Publishing Ltd in the UK. They are Senior Commissioning Manager Dr John Navas, eBooks Editorial Assistant Daniel Heatley and Production Editor Chris Benson. I'd like to thank my wife, Rongju Sun, for her support and encouragement throughout the writing of this book, and I'd like to thank my son, Guangzhi Xiao, who read my manuscript and gave me wonderful suggestions.

Author biography

Caide Xiao

Dr Caide Xiao was born in China on October 25, 1961. After obtaining his bachelor degree in Physics from Central China Normal University in 1979, he was a medical physics teacher in Yunyang Medical College, Hubei province. From 1990 to 1996, he was a postgraduate student specializing in optical biosensors in Tsinghua University, Beijing. This book is a collection of his research works around the world as a postdoctoral researcher over the last 28 years. In the book, he proposes a new design/idea for biological optical sensors. This design could dramatically simplify structures and operations of optical biosensors, and considerably raise their sensitivities.

List of symbols

γ:	a benchmark for an optical biosensor's sensitivity [degree nm^{-1}]
δ:	skin depth of light penetrating [m] or phase difference [degree]
ε_0:	vacuum permittivity, 8.8541 × 10^{-12} [F m^{-1}]
ε_r:	relative permittivity $\varepsilon_r = \varepsilon_r' + j\varepsilon_r''$
ζ:	valence of a receptor molecule (an integer)
η:	real part of a complex refractive index or a symbol for a percentage
θ:	an incident angle of a light ray [degree]
θ_{SPR}:	surface plasmon resonant angle [degree]
θ_b:	Brewster angle of a TM wave [degree]
θ_c:	critical angle of total reflection [degree]
κ:	imaginary part of a complex refractive index
λ:	wavelength of a monochromatic light wave [nm]
μ_0:	vacuum permeability, 1.2566 × 10^{-6} Henry per meter [H m^{-1}]
μ_r:	relative permeability
ν:	a damping parameter in Levenberg–Marquardt algorithm
ξ:	a damping constant in a harmonic oscillation
ρ:	ellipsometry ratio r_p/r_s or free electric charge density [C m^{-3}]
σ_{DC}:	direct current conductivity of a metal [S m^{-1}]
ϕ_o:	the initial phase angle of a harmonic wave [degree]
ϕ_p:	a TM wave's phase angle shift caused by reflection(s)
ϕ_s:	a TE wave's phase angle shift caused by reflection(s)
φ:	an outside incident angle of a SPR system or a polarizer's angle [degree]
φ_{SPR}:	an outside surface plasmon resonant angle [degree]
ψ:	an ellipsometry analyzer angle for null reflections ($\Psi + 90°$)
χ_e:	electric susceptibility of a medium: $\varepsilon_r = 1 + \chi_e$
ω_0:	natural frequency of a bound electron in an atom [rad s^{-1}]
ω_p:	plasma oscillation angular frequency [rad s^{-1}]
ω_{sp}:	cut-off frequency of a surface plasmon wave [rad s^{-1}]
ϑ:	angle position of a $\lambda/4$ plate in a null ellipsometer [degree]
Γ:	surface concentration of a biological sample on a sensing chip [ng mm^{-2}]
Δ:	a variable's increment
∇:	(del), a vector operator for gradient, divergence ($\nabla \bullet$) and curl ($\nabla\times$)
Φ:	ellipsometry phase angle ($\Phi = \phi_p - \phi_s$) [degree]
Ψ:	ellipsometry amplitude angle (analyzer principle section position) [degree]
A:	amplitude of a harmonic wave, or a unit of electric current Ampere (A)
B:	magnetic flux density [Tesla or V · s m^{-2}]
c:	light speed in a vacuum (299 792 458 m s^{-1})
C:	biological sample concentration [g mL^{-1}], or electric charge unit Coulomb
d:	thickness of thin film [nm] or a symbol for a variable's value difference
D:	electric flux density [C m^{-2}]
e:	Euler's constant 2.718 28 or absolute electron charge 1.6 × 10^{-19} C
eV:	electronvolt (1.6 × 10^{-19} J)
E:	electric field intensity [V m^{-1}]
f:	frequency of a harmonic oscillation [Hz]
f_p:	plasma oscillation frequency [Hz]
$\hbar = \frac{h}{2\pi}$:	reduced Planck constant (h = 6.626 × 10^{-34} J · s)
H:	magnetic field intensity [A m^{-1}]

I:	light intensity, or a symbol for electric current
j:	an integer or an imaginary unit for complex numbers ($j = \sqrt{-1}$)
J:	Jacobian matrix for non-linear regressions or an energy unit Joule
k:	a wavenumber vector's length $\frac{2\pi}{\lambda}$, or its horizontal component $\frac{2\pi}{\lambda}\sin(\theta)$
k_a:	a constant of association rate [M^{-1} s^{-1}] (per molar per second).
k_d:	a constant of dissociation rate [s^{-1}] (per second)
kDa:	kilo-Dalton (a unit for protein molecular weight, kg mol^{-1})
k_s:	an observable rate constant [s^{-1}] or a wavenumber vector's component
$K_A = 1/K_D$:	affinity of a ligand–receptor interaction [M^{-1}]
K_D:	a dissociation equilibrium constant [M]
ℓ:	minimum spherical diameter of a protein subunit or film thickness [nm]
Mol:	molecular molar mass [g mol^{-1} or Da]
n:	refractive index (a real number or a complex number: $n = \eta + j\kappa$)
N:	electron density in metal [m^{-3}]
p:	electric dipole moment [$C \cdot m$]
P:	polarization density [C m^{-2}]
q:	absolute electron charge 1.6×10^{-19} C
r:	a vector's length in a 3D space: $r = (x^2 + y^2 + z^2)^{1/2}$
R:	light reflectivity (energy ratio), or electric resistance [Ω]
r_p:	Fresnel coefficient for p-polarized light wave reflection
R_p:	reflectivity of p-polarized light (energy ratio)
r_s:	Fresnel coefficient for s-polarized light wave reflection
R_s:	reflectivity of s-polarized light (energy ratio)
S:	Poynting vector's length for light intensity [W m^{-2}]
t:	time [s]
T:	period of a harmonic oscillation [s]

IOP Publishing

Super Optical Biosensors

Caide Xiao

Chapter 1

A sensitivity benchmark for optical biosensors

Our lives are the results of the dynamic interactions between macromolecules inside our bodies at the right time and in the right locations. Protein is a kind of macromolecule that is used as samples in biosensors. Protein structures and their interaction affinities are introduced. Optical biosensors can be used to obtain sensorgrams about macromolecule interactions. A Mathcad program based on the Gauss–Newton algorithm demonstrates how to measure affinities by nonlinear regressions of sensorgram curves. A benchmark γ is proposed for sensitivity comparisons of optical biosensors.

Life on Earth is a miracle in the Universe. We do not know how life started on Earth, and we cannot repeat those processes to create even a single living cell from scratch. There are many hypotheses about the origin of life. In 1924, a Russia scientist A I Oparin [1] suggested that life spontaneously began in the ancient oceans on Earth. About 4.6 billion years ago, our Earth was a fireball of molten rock. After cooling down, it had an atmosphere composed of reducing gases, such as methane, ammonia, hydrogen and water vapor. Chemical reactions in the atmosphere created organic compounds which were then deposited in ancient oceans. In 1953, a US scientist S Miller [2] experimentally confirmed that electrical sparks could create amino acids and other organic compounds in a reducing atmosphere. But it is still unknown how and when these organic compounds automatically assembled into macromolecules such as DNA and protein. We know phospholipid molecules can spontaneously form vesicles as containers to hold DNA and proteins. Only when those vesicles could use energy and nutrition from their environment to multiply themselves, could they became living cells.

Each of us started from a single cell, a fertilized egg. According to DNA information in the cell, each fertilized egg differentiates into 10–100 trillion cells to construct a human body. Human beings dream of living healthily forever. But we are doomed to die. From the moment an oocyte was fertilized by a sperm to form an omnipotent stem cell, a death clock starts to count down. In a normal mammalian cell, there are 23 pairs of chromosomes (a total of 23×2). Each chromosome is a

doi:10.1088/2053-2563/ab0823ch1

spirally coiled long chain of double helix deoxyribonucleic acids (DNA) [3]. A basic unit of DNA is a nucleotide made of three components: a deoxyribose sugar, a phosphate and a nitrogenous base. The sugar and the phosphate molecules alternately link into two long chains of a chromosome. Life information is encoded by four bases on those chains: adenine (A), cytosine (C), guanine (G), and thymine (T). At each end of a human chromosome, there is a telomere: a nucleotide sequence of TTAGGG repeated approximately 2500 times [4]. Through mitosis, a cell divides to create two daughter cells to construct and maintain our bodies. But each daughter cell from a mitosis has shorter telomeres, because enzymes (protein) replicating DNA cannot do their job at the very end of each chromosome. After a limited number of cell divisions, the telomeres of chromosomes will be too short to guide cell division. Without new cells to replace dead cells to maintain our bodies, we will be sure to die.

Only if we could clone our bodies, and transcribe our memories to our clones, would it be possible for us to live forever. In this way, our lives would be beyond our bodies. Even if diseases, such as cancers, Alzheimer syndrome, and AIDS destroyed our old bodies, they would not kill us. We could even modify our genes to make our clones more handsome or beautiful, stronger, and smarter. From Dolly the sheep cloned in 1995 and many other cloned animals, we know that each nucleus of a specialized cell in our bodies contains all the information for a new baby. An unfertilized oocyte whose nucleus was removed could be used to accept the nucleus of a specialized cell to become an artificial fertilized oocyte. Enzymes inside the oocyte then read DNA information and program the oocyte to develop into a baby in a womb.

However, before we know how to transcribe our memories from one brain to another brain, our clones are useless to us. What we urgently need is to reprogram our nuclei in oocytes to create specific organs, such as hearts, livers and kidneys to replace our damaged organs to extend our lives. We have a long way to go to understand how biomacromolecules inside oocytes do their magic work. We not only have to study their 3D molecular structures by x-ray diffraction techniques, but also need to investigate their interactions in real time with biosensing techniques, because life is a result of dynamic interactions among biomacromolecules.

1.1 Affinity of biomolecular interactions

Optical biosensors are optical-electrical instruments which can be used to monitor ligand–receptor interactions in real time. A ligand is a molecule that can specifically bind with a receptor to form a complex to induce biological events. Antibody–antigen and biotin–avidin are examples for ligand–receptor pairs. The binding of insulin molecules to their receptors on the surfaces of liver cells can increase sugar absorption of liver cells to regulate blood sugar. Usually, receptors are macro biomolecules such as proteins, nucleic acids (DNA or RNA), or polysaccharides. Ligands can be macro biomolecules, as well as small inorganic molecules. Among these macro biomolecules, protein is the most important and widely used for samples in biosensing applications.

1.1.1 Amino acids and protein structures

Cells are the basic units of life. The most remarkable characteristic of life is automatic self-assembly. Twenty different kinds of amino acids inside our cells can automatically polymerize into different proteins to do their work. For example, hemoglobin molecules in our red blood cells transport oxygen and carbon dioxide in our bodies; antibodies in our blood defend our bodies from infections; insulin secreted by our pancreases can regulate our livers to take sugar from our blood. Figure 1.1 shows the chemical formulae of twenty amino acids in nine groups. Each of them has a carbon atom bonded with an amino group ($-NH_2$), a carboxyl group ($-COOH$), a hydrogen ($-H$) and a side chain (R-group). This carbon atom is named C_α to distinguish it from other carbon atoms in the carboxyl group and in the side chain. The four groups around C_α are not in a plane, but point to four corners of a triangular pyramid. We can use a camera tripod to represent a three-dimensional (3D) structure of an amino acid. If the carboxyl group is set in the camera position, and the amino group is set to be in the front leg position, then the hydrogen ($-H$) and the side chain (R-group) must be set in the two rear leg positions. In case the side chain of an amino acid is different from hydrogen, we can have an L-form or a D-form optical isomer of the amino acid. All amino acids (except glycine) exist in both

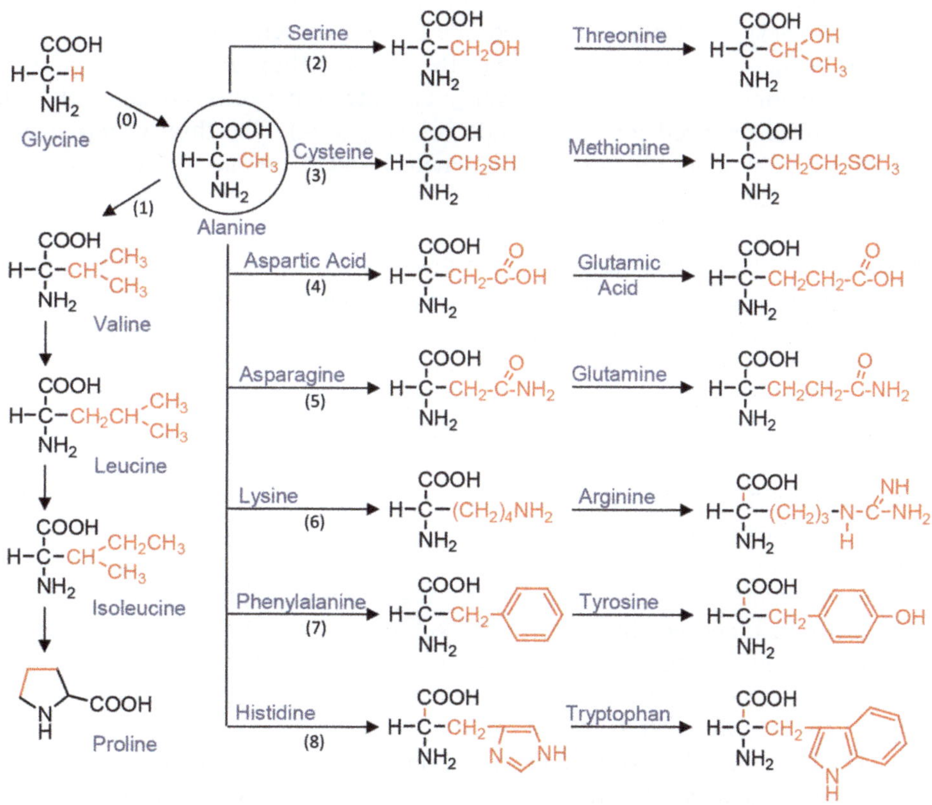

Figure 1.1. Chemical formulae of 20 amino acids in nine groups according to their side chain structures.

chiral forms, but our bodies only use L-form amino acids, whose side chain is in the rear left leg position of the tripod. Just like our left and right hands, L-form and D-form amino acids are mirror images to each other, and they rotate the polarization plane of light in opposite directions.

In biochemistry textbooks, there are many ways to group the 20 amino acids according to their physical, chemical and/or biological properties. For example, essential amino acids cannot be synthesized by our bodies whereas nonessential amino acids can be. According to the chemical groups in those side-chains, we can categorize the 20 amino acids into nine groups (figure 1.1). Glycine is the only amino acid in group 0. It is a sweet tasting amino acid whose side chain is just a proton or hydrogen (–H). Replacing the proton by a methyl group ($-CH_3$), we get alanine. Another 18 amino acids could be memorized by modifications of the methyl group in alanine. In group 1 there are five amino acids: alanine, valine, leucine, isoleucine and proline. Their hydrophobic side-chains were edited with branched methyl or ethyl groups. Proline is distinctive for its three-carbon side chain that attaches with its amino group to form a ring.

Amino acids in group 2 are serine and threonine, and their polar uncharged side-chains are related to a hydroxyl group (–OH). Serine was first found in silk by Emil Cramer in 1865, and threonine was discovered in 1936 by William Cumming Rose. Group 3 contains cysteine and methionine, and their side-chains are related to a thiol group (–SH). The polar uncharged side chain of cysteine is hydrophilic, but the side chain of methionine is hydrophobic. Because the thiol group is very active in chemical reactions, active centers of many enzymes have thiol groups. Two thiol groups could be oxidized into a disulfide bridge (–s–s–). The thiol group strongly reacts with a gold atom [5] to form a covalent bond. Cysteine molecules could be used to block unoccupied gold surfaces on biosensing chips to reduce unspecific interactions.

Aspartic acid and glutamic acid are in group 4, and each of them has a second carboxyl group (–COOH) in their side-chains which is negatively charged in the physiological condition of our bodies. The polar uncharged side-chains of asparagine and glutamine in group 5 are related to an amide group ($-CONH_2$). Asparagine was first discovered in 1806 from asparagus. After eating asparagus, some of us will have urine with a strange smell. Lysine and arginine are in group 6, and their positive charged side-chains are related to another amino group ($-NH_2$). In active centers of many enzymes there are lysine residues, because the amino group is very active in biochemistry reactions.

The two amino acids in group 7 are phenylalanine and tyrosine. Their side hydrophobic chains are related to a benzyl ring. Phenylalanine is famous because it is related to an inherited disorder: phenylketonuria (PKU). A mutated gene of a PKU patient cannot make a normal enzyme to metabolize phenylalanine from food. The concentration of phenylalanine accumulates to a toxic level in the patient whose urine has a particular smell. Histidine and tryptophan are two amino acids in the last group. The first one is hydrophilic for its imidazole side chain with a positive charge in the physiological condition of our bodies, and the second one hydrophobic. Histidine residues in enzymes are active in their catalytic sites.

Each protein has at least one linearly polymerized amino acid polypeptide chain in which two neighboring amino acids are linked by a peptide bond discovered by Linus Pauling [6]. In a dehydration synthesis reaction, a carboxyl group linked with C_α in one amino acid reacts with an amino group linked with C_α of another amino acid. The dehydration reaction creates a peptide bond (–CO–NH–) between the two amino acids, and releases a water molecule. The primary structure of a protein is its linear sequence of amino acid residues in its polypeptide chains. For example, bovine insulin has two linear polypeptide chains named chain A and chain B. They have 21 and 30 amino acid residues respectively, and are linked by two disulfide bridges. The sequence of the chain A is: Gly-Ile-Val-Glu-Gln-Cys-Cys-Ala-Ser-Val-Cys-Ser-Leu-Tyr-Gln-Leu-Glu-Asn-Tyr-Cys-Asn. It is written from one chain end with a free amine group called the 'N-terminus'. The other end of the chain with a free carboxyl group linked to Cα is called the 'C-terminus'. The amino acid sequences of a protein are genetically encoded by a gene. Three DNA bases (a triplet) in a gene codes an amino acid in protein.

Because of the hydrophilic and hydrophobic properties of side-chains of amino acid residues, the linear chain in water can spontaneously fold into a secondary structure (α-helixes, β sheets and random folding) and a tertiary structure (3D structure). The minimum diameter of a protein sphere is estimated by a formula (equation (1.1)). For protein, we use nanometer (1 nm = 10^{-9} m) as their size unit, and kilo-Dalton (kDa) as their molar mass unit (1 Da = 1 g mol^{-1}). One molar mass (*Mol*) is the total mass of 6.022×10^{23} (Avogadro's number) molecules. For water and hemoglobin, their molar mass is 18 and 64 000 Da respectively. Given the molecular mass value of a protein in kilo-Dalton, we can calculate its minimum spherical diameters in nanometers [7].

$$\ell = 1.32\sqrt[3]{Mol} \tag{1.1}$$

The calculation is based on all proteins having almost the same density: 1.37 [g cm^{-3}]. For example, bovine serum albumin (BSA), a serum albumin protein in cows' blood has a molar mass of 66.5 kDa, its minimum diameter is about 5.4 nm. The above formula cannot be used for proteins with more than one subunit.

1.1.2 Antigen–antibody interaction on the surface of gold

With more than one subunit, a protein such as hemoglobin has a quaternary structure: the combination and arrangement of multiple subunits. Immunoglobulin G (IgG) in our blood is a kind of antibody. Each IgG has two identical heavy chains (~50 kDa) and two identical light chains (~25 kDa). The four subunits automatically assemble into a 3D structure in a 'Y' shape, shown in figure 1.2(C). The fork ends of the 'Y' are two identical binding sites for specific antigens. They are called Fab regions, because they are variable parts that specifically target what the antibody should bind to. The tail part of the 'Y' is the Fc region. It can bind with protein A, a surface protein (42 kDa) extracted from bacteria *Staphylococcus aureus*. According to the formula (1.1), the diameter of protein A is about 4.6 nm. Figure 1.2 shows antigen–antibody interactions detected by a surface plasmon resonance (SPR)

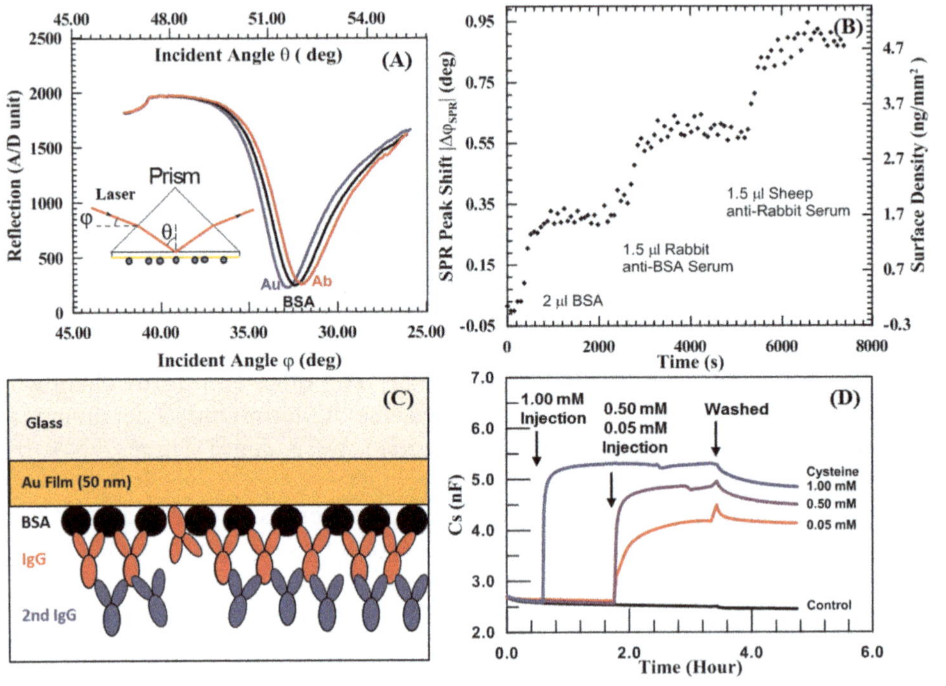

Figure 1.2. (A) Surface plasmon resonance spectra of a gold film, BSA and rabbit anti-BSA IgG. (B) Relationships between $|\Delta\varphi_{SPR}|$ and time during BSA adsorption on a gold film, antibodies binding to BSA, and the second antibodies binding with the first antibodies. (C) A sketch of antigen–antibody interactions about BSA, IgG and second IgG (not to scale). (D) Amino acid cysteine binding onto gold electrodes monitored by electric capacitance changing at 4 kHz.

biosensor [8] shown in chapter 6. The antigen sample is bovine serum albumin (BSA), the antibody sample is rabbit anti-BSA serum, and a second antibody sample is sheep anti-rabbit serum. Dry powders of these protein samples were dissolved in a phosphate-buffered saline (PBS, pH 7.2) to 1.0 mg mL^{-1} respectively. With a SPR biosensor, antigen–antibody interactions can be detected in 30 minutes with samples of only a few microliters. Liedberg [9] and his colleagues in 1983 first reported the novel application of SPR.

The inset of figure 1.2(A) shows a SPR sensing chip: a coverslip with a gold film and a BSA film. The glass side of the coverslip was attached to the base of a prism by a drop of refractive index matching oil for microscopy. A piece of Plexiglas with a rubber O-ring was pressed onto the gold film to form a chamber with a volume of around 250 µl. After the chamber of the SPR biosensor (see chapter 6) was filled with 150 µl PBS, a SPR spectrum was obtained (the blue curve) shown in figure 1.2(A). 1.5 µl BSA was then injected into the chamber. After 30 minutes, the chamber was washed and refilled with 150 µl PBS, and then the second SPR spectrum was measured (the black curve). Then rabbit anti-BSA serum 1.5 µl was injected into the chamber. Also after 30 minutes, the chamber was washed and refilled with 150 µl PBS, and the third SPR spectrum was obtained (the orange curve). The BSA/IgG/anti-IgG monolayers over the gold surface shifted the SPR spectra from left to right. A SPR angle is defined as the light incident angle corresponding to the lowest reflectivity of a SPR spectrum. There is an outside SPR angle φ_{SPR} and an inside SPR angle θ_{SPR}. Figure 1.2(B) shows the absolute shift of φ_{SPR} with time during three processes: antigen (BSA) molecules binding onto the gold surface, antibody (rabbit anti-BSA serum) molecules specifically binding with BSA, and the second antibody (sheep anti-rabbit serum) molecules binding with the first antibodies. Sometimes in medical applications of SPR biosensors, we have to use antibody samples to detect antigens. But antibody molecules will not react with their antigens if antibodies directly bind onto gold surfaces. To overcome this problem, we could use protein A to immobilize antibodies on the gold surface. Protein molecules have orientations. Molecules of protein A spontaneously binding to a gold film can prevent IgG molecules from contacting with the gold surface. Since only the Fc end of each IgG molecule binds with protein A, the antibody molecules are properly aligned for their functions.

Antibodies specifically bound to their antigens can be partially washed away using citric acid saline (~pH 2.0, 0.85% NaCl) because of non-specific binding. As in the schematic shown in figure 1.2(C), antibodies can bind to the gold surface if BSA is not homogeneously distributed on the gold surface in a spontaneous adsorption. Biological samples assemble over the gold surface can be seen directly with a SPR microscope [10]. Any protein with cysteine residual(s) would strongly bind to the gold surface through a chemical reaction between the –SH group and a gold atom. The chemical reaction creates a covalent bond between the sulfur atom and the gold atom [11, 12]. After antigen molecules are immobilized on the gold surface of a sensing chip, we can use cysteine to cover the exposed gold surface to reduce non-specific binding of antibodies. It is not possible to use SPR biosensors to detect the self-assembly monolayer (SAM) of cysteine on the gold surface, because that cysteine SAM is too thin for SPR biosensors. However, by measuring the electric capacitance change of gold electrodes in 0.85% NaCl saline, it is easy to detect the cysteine SAM. Figure 1.2(D) shows capacitance C_s changing with time after L-cysteine (6 μl) solutions were injected into wells each with 600 μl 0.85% NaCl saline [13]. In each well there is a small gold electrode (ϕ 0.25 mm) and a larger gold electrode (2×9 mm^2). The concentrations (1.00, 0.50 and 0.05 mM) shown in the figure were the final cysteine concentrations in each well respectively. With two bigger gold electrodes (diameter 3 mm) in the saline, the capacitance change caused by the cysteine SAM could be as high as 450 nF. Cysteine with an active thiol group is particularly useful for electrode modifications.

1.1.3 Equilibrium dialysis for antigen–antibody affinity measurements

Antigen–antibody interactions are examples of ligand–receptor interactions. We not only need to know whether a ligand binds with its receptor, but also want to measure the strength of the bond. When different ligands with almost similar chemical structure compete to bind with the same kind of receptor molecules, we should know the difference of their binding strengths or affinities. In the pharmaceutical industry, researchers modify chemical structures to invent drugs. The invention of an antibacterial chemical drug Prontosil by Gerhard Domagk in the 1930s is an example [14]. One Prontosil molecule in the human body can be metabolized into an azo-dye molecule and a sulfonamide molecule. The second one is very similar to p-amino benzoic acid (pABA), an essential compound for the synthesis of folate (vitamin B9) which is needed in bacterial growth and reproduction. In folate synthesis, pABA should bind with an enzyme dihydropteroate synthase (DHPS). With a similar chemical structure, sulfonamide can 'fool' DHPS, and irreversibly bind with DHPS to block folate synthesis. Without folate supply, bacteria will be starved to death.

In 1894 Emil Fischer postulated a lock-and-key model [15] to explain ligand–receptor interactions. A protein receptor's linear polymerized amino acid chains automatically fold into a unique geometric 3D shape, so that its active sites are complementary to the geometric shape of its ligand molecule(s). Anfinsen [16] proposed a thermodynamic hypothesis to explain the spontaneous folding and automatic assembling of protein to a minimum energy configuration. Amino acid residues with hydrophobic side-chains tend to hide themselves inside the protein to avoid contact with water, and other amino acid residues with hydrophilic side-chains tend to be exposed on the protein's surface. The lock-and-key model specifies that receptors can only selectively bind with very special molecules with a complimentary structure. Actually, biomolecules are flexible, not as rigid as locks and keys. A new theory of induced fit coined by Daniel E Koshland [17] in 1959 postulates that, due to protein segments' flexibility, when a ligand is close to its enzyme receptor, the enzyme protein can be induced to change its shape to better match its approaching ligand.

The affinity of a ligand to its receptor can be directly measured by equilibrium dialysis. In an association reaction shown by the right facing arrow in equation (1.2), a ligand molecule binds with a receptor molecule to form a complex of the ligand–receptor pair. The reaction stoichiometry is 1:1 which is true for many kinds of biological binding events. The increasing rate of the complex concentration $[C]$ is proportional to the product of the free ligand concentration $[L]$ and the free receptor concentration $[R]$ at a given time in the system: $k_a \cdot [L] \cdot [R]$, where k_a is an association rate constant with a unit of $[\mathrm{M}^{-1}\ \mathrm{s}^{-1}]$ (per molar per second). The association reaction is a second order reaction.

$$\text{Ligand} + \text{Receptor} \underset{k_d}{\overset{k_a}{\rightleftharpoons}} \text{complex} \tag{1.2}$$

Let time be zero ($t = 0$) when the ligand–receptor interaction starts and the initial ligand and receptor concentrations be $[L_0]$ and $[R_0]$ respectively. While the complex concentration is increasing, concentrations of the free ligand and free receptor molecules will decrease. With $t \geqslant 0$, we have two equations: $[L] = [L_0] - [C]$ and $[R] = [R_0] - [C]$. The value of k_a is determined by the collision rates between ligand and receptor molecules.

In a reverse reaction shown by the left facing arrow in equation (1.2), one molecule of the complex could dissociate into a ligand molecule and a receptor molecule. The decreasing rate of the complex is only proportional to the complex concentration $k_d \cdot [C]$, where the parameter k_d is defined as a dissociation rate constant with units of [s^{-1}] (per second). The dissociation reaction is a first-order reaction.

According to the law of mass interaction, at an equilibrium condition of the ligand–receptor interaction, the increasing rate and the decreasing rate of the complex should be the same: $k_a \cdot [L] \cdot [R] = k_d \cdot [C]$. From this equation, we obtain the dissociation equilibrium constant of the ligand–receptor interaction with a unit of molar concentration [M]:

$$K_D = \frac{[L] \cdot [R]}{[C]} \tag{1.3}$$

The reciprocal of K_D is defined as the affinity (K_A) of the ligand–receptor interaction with a unit of [M^{-1}]:

$$K_A = \frac{[C]}{[L] \cdot [R]} \tag{1.4}$$

To measure the affinity of a ligand–receptor interaction, we have to precisely measure concentrations of reactants using techniques such as chemical assay or optical absorption assay. In addition, all the measurements must not disturb the equilibrium of the reaction.

A simple technique called equilibrium dialysis [18] can be used to measure affinities of ligand–receptor interactions. It is based on the huge size difference between the two reactants. The key part of the equilibrium dialysis technique is a semipermeable membrane made of cellulose that partitions a solution container. There are holes on the membrane. The holes are big enough for the ligand molecules to pass through freely, but block the receptor molecules from passing through. According to the formula (1.1), the diameter of the holes on the membrane should be less than a few nanometers. A simplified example of equilibrium dialysis assay is shown in figure 1.3, in which a semipermeable membrane separates a chamber filled with water. At the start of the assay ($t = 0$), 6 μM ($[R_0]$) of receptor molecules are in the left half of the chamber, and 10 μM ($[L_0]$) of ligand molecules are in the right half

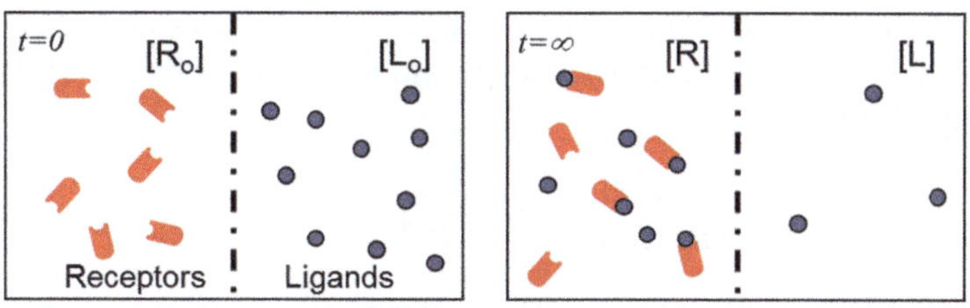

Figure 1.3. Using the equilibrium dialysis technique to measure the affinity of a ligand–receptor interaction with a semipermeable membrane shown by the dashed line in the middle of a water chamber.

of the chamber. In most cases, the reaction reaches equilibrium in a few minutes. If the ligand molecules are not specific for the receptor molecules, they would not bind with receptors to form complexes, so the free ligand concentrations on both sides of the membrane should be the same; otherwise, the left side of the chamber would have a higher ligand concentration, as shown on the right of figure 1.3.

Should we find the free ligand concentration on the right side of the membrane is $[L] = 3\ \mu M$, we are sure that the free ligand concentration on the left side of the membrane is also 3 μM; the rest of the 4 μM ligand molecules on the left side must bind with 4 μM receptor molecules to form complex molecules with a concentration $[C] = 4\ \mu M$; and the free receptor concentration in the left side of the chamber would be $[R] = 2\ \mu M$. From the formula (1.4), we get the affinity of the ligand–receptor interaction: $K_A = 0.67\ \mu M^{-1}$, and the dissociation equilibrium constant $K_D = 1.5\ \mu M$. The smaller K_D values indicate higher affinity of a ligand–receptor interaction. In an extreme condition, the K_D value is equal to zero if $[L] = 0$ or/and $[R] = 0$ at the end of a ligand–receptor reaction. Usually [19], for a lower affinity interaction, the K_D value could be as high as 1 μM, and for a higher affinity interaction the K_D value could be as low as $10^{-3}\ \mu M$. Avidin is a tetrameric protein extracted from egg white. Each of its four subunits can bind with a biotin with a K_D value of approximately $10^{-12}\ \mu M$. The avidin/biotin interaction can form one of the strongest known non-covalent bonds [20] which could be used to make biochips for biosensors. Logically, the bigger the K_A value, the higher the affinity of a ligand–receptor is. The affinity of carbon monoxide to hemoglobin is about 200 times as strong as that of oxygen [21]. The unit μM^{-1} of affinity K_A is easy to understand by converting it into volume/molecule or the radius of effective interaction between a ligand and a receptor.

Sometimes, one receptor molecule can specifically bind with more than one ligand molecule. The number of binding sites of a receptor molecule is the receptor's valence, an integer ζ ($\geqslant 1$). The valence of an antibody IgG molecule and an avidin molecule are two and four respectively. Using the Scatchard equation [22], we can analyze equilibrium dialysis data. In a set of equilibrium dialysis experiments with a fixed concentration $[R_0]$ of a receptor, we vary the ligand concentration $[L_0]$ to obtain a receptor's valence ζ and the affinity of a binding site to a ligand. By using ligands with radioactive labels in these experiments, we can measure radioactive intensities on both sides of the dialysis chamber to calculate the free ligand concentration $[L]$ as well as the bound receptor binding site concentration $[C] = [L_b]$, which is the radioactive intensity difference of the receptor chamber and the ligand chamber shown in figure 1.3.

The free binding site concentration of the receptor with valence ζ can be deduced as $[R] = \zeta[R_0] - [L_b]$. By taking these parameters into formula (1.4), we get the Scatchard equation

$$\frac{1}{[L]}\frac{[L_b]}{[R_0]} = -K_A\frac{[L_b]}{[R_0]} + \zeta K_A \tag{1.5}$$

By repeating the equilibrium dialysis experiments while only varying the ligand concentration $[L_0]$, we can draw a Scatchard plot with $[L_b]/[R_0]$ as its x-variable and

$[L_b]/[R_0]/[L]$ as its y-variable. If the affinity of each binding site of a receptor is the same, we will have a straight line with a slope $-K_A$ and a y-intercept ζK_A on the Scatchard plot. The absolute slope value of the line is the affinity of each binding site of the receptor to a ligand, and the ratio of the intercept to the affinity is equal to the valence ζ. The overall strength of the receptor binding to its ligand molecules is called avidity: ζK_A. A hemoglobin molecule has four subunits and each of them can reversibly bind with an oxygen molecule. After the first O_2 binds with a subunit in a hemoglobin molecule, the protein changes its shape to increase the affinities of its remaining subunits to O_2 molecules. In that case, the Scatchard curve of hemoglobin is not a straight line.

1.1.4 Kinetic binding assay

We do not need to label our sample molecules with radiative elements or fluorescent dyes when using optical biosensors to perform kinetic binding assays for affinity measurements. We can also quickly obtain assay results with far fewer samples by using optical biosensors although they are more expensive. The samples of optical biosensors could be proteins, nucleic acids, polysaccharides or lipids. Biomolecules are immobilized on the surface of a solid material such as gold to make sensing chips. The immobilized biomolecules on a chip are called receptors which can specifically recognize and bind with their ligands. Usually at least one biochip is installed in a wet chamber of an optical biosensor. After ligand molecules are injected into the chamber, they will specifically bind with the receptors to form a second layer on the top of the receptors as shown in figure 1.4. The ligand film has a thickness which is equal to the diameter of the ligand, and the surface concentration of the ligand molecules would increase from zero to a level of saturation during a binding process. The film of biological material with a thickness of only a few nanometers can modulate the incident light in a unique way of a special sensing technique. The light intensity, phase, wavelength, or incident angles could all be modulated. Biological signals carried by an output light in the optical system are

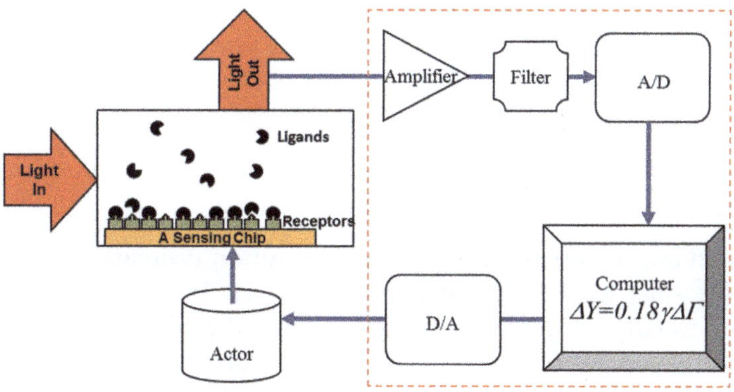

Figure 1.4. A block diagram of a universal optical biosensor. Reactions of ligands with receptors on a sensing chip modulate the input light to generate electric signals which are amplified, filtered and digitalized in a computer. An analog signal is also generated by the computer to an actor for an angle scan process.

converted into electrical signals by light detectors. Then in a computer, the electrical signals are amplified, filtered, digitalized and recorded. Analog signals may need to be sent out by the computer to actors for angle scan or wavelength scan processes.

The initial concentrations of the two reactants are $[L_0]$ and $[R_0]$ respectively in a biosensor chamber and the valence of the receptor is supposed to be one. While a receptor molecule binds to a ligand molecule with an association rate constant k_a to form a complex, the complex can also disassociate into a receptor molecule and a ligand molecule with dissociation rate constant k_d. The complex concentration changes with time before the reactions reach equilibrium

$$\frac{d[C]}{dt} = k_a[R][L] - k_d[C] = k_a([R_0] - [C])([L_0] - [C]) - k_d[C] \tag{1.6}$$

Since the complex changing rate is proportional to the square of the complex concentration, the reaction of the ligand–receptor is a second order reaction. Mathematically there are no explicit solutions for equation (1.6), and practically it is difficult to measure concentrations of the two reactants simultaneously. We can circumvent these problems by a method called 'pseudo-first-order approximation'. By supplying the ligand concentration to the biosensor chamber in a great excess to the receptor concentration, we can satisfy $[L_0] \gg [R_0]$, so that the amount of ligand molecules consumed by the experiment is negligible compared to the initial ligand concentration: $[L_0] - [C] \approx [L_0]$

$$\frac{d[C]}{dt} = k_a[L_0]([R_0] - [C]) - k_d[C] \tag{1.7}$$

The solution of the above pseudo-first-order equation is similar to that of an equation about a resistor and a capacitor connected in series charged by a battery

$$[C] = \frac{[L_0][R_0]}{[L_0] + k_d/k_a}\left(1 - e^{-(k_a[L_0]+k_d)t}\right) \tag{1.8}$$

At time $t = \tau$, the complex concentration in the chamber would be increased to $[C_\tau]$, and the chamber is then washed by water (or buffer) in an elution process in which there are possibilities for complex molecules to disassociate into ligand and receptor molecules. The decreasing rate of the complex is proportional to the concentration of the complex: $\frac{d[C]}{dt} = -k_d[C]$ and its solution is:

$$[C] = [C_\tau]e^{-k_d(t-\tau)} \tag{1.9}$$

From the solution, we see the complex concentration decreases exponentially with time from the start of the elution process ($t = \tau$). In optical biosensors, ligand molecules are usually adsorbed on the flat surface of a biochip or a sensing chip to form a ligand layer over a receptor film. The surface concentration (Γ) of the complex is equal to its volume concentration $[C]$ times its thickness (ℓ). The thickness of the complex film should be equal to the sum of the receptor thickness ℓ_R and the ligand thickness ℓ_L. The units of $[C]$, Γ and ℓ are [g mL^{-1}], [ng mm^{-2}] and [nm] respectively. The volume concentration of a sample adsorbed on the biochip surface is much higher than the volume concentration of a sample in the wet chamber. A study by Klintström showed a 10 000× difference [23]. Because samples are enriched on the surface of biochips during ligand–receptor interaction, we only focus on the sample volume concentration on the biochip surface. The complex concentration change ($\Delta[C] = [C] - 0$) will also change the optical refractive index of the complex film

$$\Delta\Gamma = \Delta[C] \cdot \ell = \left(\frac{\partial[C]}{\partial n}\right)\Delta n \cdot \ell = \left(\frac{\partial n}{\partial[C]}\right)^{-1} \Delta(n \cdot \ell) \tag{1.10}$$

From the studies of Chincholi [24] and Feijter [25], $\frac{\partial n}{\partial[C]}$ can be treated as a constant: 0.18 [mL g^{-1}] for biological samples. The product of a film's geometric thickness ℓ and its refractive index n is the effective optical thickness of the film. If a biosensor's measurable parameter Y is proportional to the effective optical thickness of its sample film on biochips, we can define a benchmark γ for the optical biosensor's sensitivity. For a surface plasmon resonance (SPR) biosensor, the measurable parameter Y is a shift of its incident light angle for surface plasmon resonance. The unit of a SPR biosensor's benchmark γ should be [deg nm^{-1}]. If the measurable parameter Y of a biosensor is wavelength, the benchmark is a dimensionless parameter

$$\Delta Y = \gamma \cdot \Delta(n \cdot \ell) \tag{1.11}$$

A good biosensor should have a bigger benchmark value and a wider linear response range than its sample effective optical thickness. By combining equations (1.10) and (1.11), we establish a link between the change of a biosensor's measurable parameter Y and the change of the surface concentration of its sample

$$\Delta Y = 0.18 \cdot \gamma \cdot \Delta\Gamma \tag{1.12}$$

With the above formula, we can use the universal biosensor to precisely measure the surface concentration of sample molecules on a sensing chip. In the beginning of an affinity assay measured with the universal optical biosensor, we install a sensing chip with a virgin gold surface, and inject receptor molecules into a wet chamber of the biosensor. After these molecules automatically form a monolayer on the gold surface, the chamber is washed and refilled with water or buffer to record ΔY_R, the change of the biosensor's parameter Y corresponding to the receptor film on the gold surface. The volume concentration of the receptor molecules in the monolayer on the sensing chip should be: $[R_0] = \frac{\Delta\Gamma_R}{\ell_R} = \frac{\Delta Y_R}{0.18\gamma}\frac{1}{\ell_R}$. Then ligand molecules are injected into the chamber and the ligand concentration in the chamber is $[L_0] \gg [R_0]$. Shortly after, there will be a ligand monolayer over the receptor monolayer. At a time after the ligand–receptor reactions begin, the change of the biosensor's parameter based on Y_R should be ΔY. Since the initial complex concentration is zero, we have $[C] = \frac{\Delta\Gamma}{\ell_R + \ell_L} = \frac{\Delta(\Gamma_R + \Gamma_L)}{\ell_R + \ell_L} = \frac{\Delta\Gamma_L}{\ell_R + \ell_L} = \frac{1}{\ell_R + \ell_L}\frac{\Delta Y}{0.18\gamma}$ for a film over the gold surface. By taking $[R_0]$ and $[C]$ into the formulae (1.8) and (1.9), we obtain a theory sensorgram curve of the universal biosensor's sensorgram

$$\Delta Y = \begin{cases} A_\infty(1 - e^{-k_s t}) & (0 \leqslant t < \tau) \\ A_\tau e^{-k_d(t-\tau)} & (t \geqslant \tau) \end{cases} \tag{1.13}$$

In the above functions for a sensorgram curve with two segments, the parameter k_s is 'an observable rate constant', and the parameter A_∞ is equal to the maximum change of Y at equilibrium for a ligand concentration $[L_0]$

$$k_s = k_a[L_0] + k_d \tag{1.14}$$

$$A_\infty = \frac{\ell_L + \ell_R}{\ell_R}\frac{[L_0]}{[L_0] + K_D}\Delta Y_R \tag{1.15}$$

$$A_\tau = A_\infty[1 - e^{-k_s\tau}] \tag{1.16}$$

Figure 1.5 is a surface plasmon resonance (SPR) sensorgram [26] about interactions of bovine carbonic anhydrase II to benzenesulfonamide chemically linked to a self-assembled monolayer (SAM) of alkanethiolates on a gold surface.

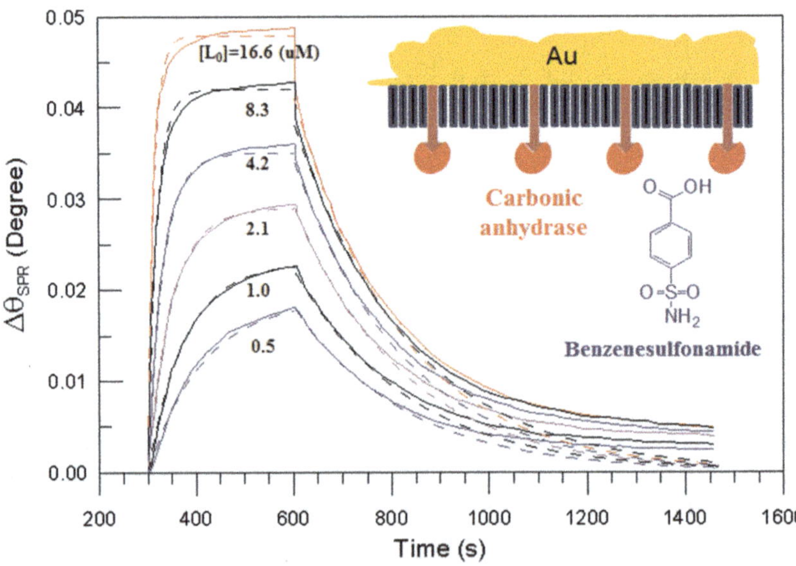

Figure 1.5. A sensorgram about interactions between bovine carbonic anhydrase and its inhibitor benzenesulfonamide. The solid curves are based on digitalized data of figure 12(A) in Lahiri's paper [26], and the dashed lines came from nonlinear least squares fits. The inset is about benzenesulfonamide receptors immobilized to a self-assembled monolayer (SAM) of alkanethiolates on a gold surface. Reprinted with permission from [26]. Copyright (1999) American Chemical Society.

The carbonic anhydrase (29 kDa) is an enzyme that rapidly catalyzes the interconversion between carbon dioxide and carbonic acid: $CO_2 + H_2O \rightleftharpoons H_2CO_3$. This enzyme has two functions: maintaining the pH balance in blood and helping red blood cells to transport carbon dioxide out of tissues. Carbon dioxide and oxygen molecules can freely cross blood cell membranes by a process of simple diffusion, but carbonic acid cannot. After CO_2 molecules in tissues diffuse into red blood cells, the carbonic anhydrase quickly converts CO_2 into H_2CO_3. In lungs, the enzyme reversely converts H_2CO_3 into CO_2 and water. The CO_2 molecules then diffuse out of the red blood cells into exhaled air. A molecule of the enzyme has an active site in a pocket in its 3D structure. There are four histidine side-chains in the pocket to catalyze their substrate reactions. As an inhibitor of the enzyme, benzenesulfonamide [27] can be used as an efficient tumor growth inhibitor [28]. The sensorgram was obtained by injecting carbonic anhydrase (16.6, 8.3, 4.2, 2.1, 1.0, and 0.5 μM) over surfaces of sensing chips with multiple channels. On the gold surfaces of the sensing chips, there were mixed self-assembled monolayers of alkanethiolates. The shorter and longer thiol molecules are $HS(CH_2)_{11}$ $(OCH_2CH_2)_3OH$ and $HS(CH_2)_{11}$ $(OCH_2CH_2)_6OCH_2O_2H$ respectively. The molar ratio of the two thiols is 100:1, and the inhibitor molecule is chemically linked to a longer thiol. By nonlinear regressions of the six association curves and the disassociations curves, Lahiri's paper reported values of k_a, k_d, K_D and K_A are 0.0094 μM^{-1} s^{-1}, 0.0054 s^{-1}, 0.57 μM and 1.74 μM^{-1} respectively.

1.2 Nonlinear regression of sensorgrams for affinity measurements

Any sensorgram coming from an optical biosensor has affinity information hidden in all data points on sensorgram curves. On a segment of a sensorgram curve, there are $(N + 1)$ data points measured by the optical biosensor: $\{(t_i, y_i), i = 0, 1, \ldots, N\}$. By using the Gauss–Newton algorithm for nonlinear least squares fitting, we can extract affinity information about the ligand–receptor interactions.

1.2.1 The principle of the Gauss–Newton algorithm

A theoretical function $y = f(t, X)$ is needed to model a segment of a sensorgram curve with $(N + 1)$ points, where $X = [\chi_0, \chi_1, \ldots, \chi_M]^T$ is a matrix or vector, and N and M are integers with $N \geqslant M$. The superscript T of the vector X indicates a mathematical operation called 'matrix transpose'. Using Gauss–Newton algorithm regression, we must construct a function $S(X)$ which is a half of the sum of the squared residuals between the model values $f_i = f(t_i, X)$ and the observed values y_i at all test points ($i = 0, 1, \ldots, N$)

$$S(X) = \frac{1}{2}\sum_{i=0}^{N}[f_i - y_i]^2 = \frac{1}{2}R^T \cdot R \tag{1.17}$$

where $R = [r_0, r_1, \ldots r_N]^T = [F(X) - Y]$, $F(X) = [f(t_0, X), f(t_1, X), \ldots f(t_N, X)]^T$, and $Y = [y_0, y_1, \ldots y_N]^T$ are three vectors respectively. An optimal parameter vector

X that minimize $S(X)$ comes from a condition by which the gradient of $S(X)$ to X is equal to zero [29]

$$\nabla S(X) = \left[\frac{\partial S(X)}{\partial \chi_0}, \frac{\partial S(X)}{\partial \chi_1}, \ldots, \frac{\partial S(X)}{\partial \chi_M} \right]^{\mathrm{T}} = J^T \cdot R = 0 \tag{1.18}$$

where the symbol 'J' represents a Jacobian matrix with $J_{ij} = \frac{\partial f_i}{\partial \chi_j}$, ($i = 0,1,\ldots,N$, and $j = 0,1,\ldots,M$). Usually, it is not easy to solve the equation $\nabla S(X) = 0$, but we can use the Gauss–Newton algorithm [30] to obtain the optimal parameter matrix from an initial vector or matrix X_0 by iterations

$$X_{k+1} = X_k - [J^T \cdot J]^{-1} \cdot J^T \cdot R \tag{1.19}$$

where the superscript '-1' indicates a mathematical operation called 'matrix inverse'. Formula (1.19) is a modified Newton method [31] which could be used to find the roots of an equation with only one variable by iterations from an initial guessed value: $x_{k+1} = x_k - f(x_k)/f'(x_k)$. To get the formula (1.19), we have to set X in equation (1.18) to X_{k+1} and then replace $R(X_{k+1})$ by a Taylor expansion at X_k: $R(X_{k+1}) \approx R(X_k) + J \cdot \Delta X$. For a quick convergence of the Gauss–Newton algorithm, we have to choose an initial X_0 as close as possible to the optimal X so that the higher ranks of ΔX in the Taylor expansion of $R(X_{k+1})$ can be ignored. In this way, we have $J^T \cdot (R + J \cdot \Delta X) = 0$ in equation (1.18), and from which equation (1.19) can be derived.

For an association curve shown in figure 1.5, we need to model the sensorgram data ($300 \leqslant t \leqslant 600$) to a nonlinear function $y = \alpha(1 - e^{-\beta t})$ with only two parameters (α and β) to be determined. At each data point of the association curve on a sensorgram, the residual between the model value and the observed value is $r_i = \alpha(1 - e^{-\beta t_i}) - y_i$. We then have the Jacobian matrix shown below from partial derivatives of the nonlinear function to the two parameters respectively: $\frac{\partial}{\partial \alpha} f_i = 1 - e^{-\beta t_i}$ and $\frac{\partial}{\partial \beta} f_i = \alpha t_i e^{-\beta t_i}$. The residual matrix is also explicitly shown on the right side of the Jacobian matrix, where $N = 79$

$$J = \begin{bmatrix} (1 - e^{-\beta t_0}) & \alpha t_0 e^{-\beta t_0} \\ (1 - e^{-\beta t_1}) & \alpha t_1 e^{-\beta t_1} \\ \cdots & \cdots \\ (1 - e^{-\beta t_N}) & \alpha t_N e^{-\beta t_N} \end{bmatrix} \quad R = \begin{bmatrix} \alpha(1 - e^{-\beta t_0}) - y_0 \\ \alpha(1 - e^{-\beta t_1}) - y_1 \\ \cdots \\ \alpha(1 - e^{-\beta t_N}) - y_N \end{bmatrix} \tag{1.20}$$

1.2.2 Introduction of MathCAD software for nonlinear regressions

The mathematics software Mathcad® from Parametric Technology Corporation is used in this book for numeric simulations of optical biosensors and data analysis. The unique characteristic of this software is that all physics formulae in textbooks can be directly used in their mathematical forms. It is easy to read and write Mathcad programs for numerical simulations of complex physical phenomena.

When Mathcad is launched on a computer with a Microsoft Windows operating system, its interface is a blank page with standard Windows menus and Mathematics toolbars at the top of its interface. A small red cross will be on the top-left corner of the blank page, and it will be moved anywhere on the page where you left-click your mouse. The location of the red cross is where you type your code. You can use Mathcad as a calculator for simple calculations. However, Mathcad is more powerful with variables and sub-programs.

On the interface, if you type an alphabetic letter or word followed by a colon (:), Mathcad will treat the letter or word as a variable and automatically change the colon into (:=). For example, after you type x:5, you will see x:=5 on the interface. A variable is a container to store data, a function, or a string. The data can be a scalar, a matrix or even a table read from a Microsoft Excel file. If a matrix B has m (>1) rows and 1 column, we can say matrix B is a vector with m dimensions. The one column matrix B mentioned above could be written as: $B = [b_0, b_1,\ldots b_{m-1}]^T$ to save page space. Variable names are case, font, size, and style sensitive. A variable name can only start with an English or Greek letter and not with a number from 0 to 9. Greek alphabets can be input from a Greek toolbar on the interface or directly from your English keyboard. For example, θ can be input by typing q, and then while holding Ctrl key pressing g (q+Ctrl+g).

After you type any numbers and any alphabetic letters and then press the spacebar, Mathcad treats the input as a text field with a black rectangular frame. You can type as many words as you want in the text field to tell your readers what you want your code to do. The text font size, color and style can also be changed with the formatting toolbars on the interface. Mathcad evaluates your codes from top to bottom and left to right of the interface. Any variable being used for calculations must be assigned with a number on the top and/or left of the Mathcad programming interface. Local variables used in a sub-program or function defined by users must be assigned values when that function is called. There is a limitation in Mathcad programs about variable names. Each optical parameter in biosensors is given a name for a variable in mathematic formulae. Any of these variable names is usually constructed by an alphabet and a digital number. The digital number following its alphabet is in subscript for elegance of those formulae. For example, n_1, n_2 and n_3 are variable names for media refractive indices, d_2 is a variable name for a film thickness of medium 2, and θ_0 is a variable name for light incident angle in medium 0. But for local variables of Mathcad sub-programs, the digital number in each local variable name cannot be in subscript, so those names have to be in another form: n1, n2, n3, d2, θ0, etc. The two forms of each variable are alternately used in the book and they are the same.

A Mathcad program for a nonlinear least squares fitting based on the Gauss–Newton algorithm for sensorgrams is shown in figure 1.6. Toolbars are hidden to save space. The first two lines of the program are text regions. A Microsoft Excel data file for the sensorgram shown in figure 1.5 was read into the Mathcad program by using menus 'Insert → Data → File Input' on the third line. The Excel data file input into the program becomes a big matrix with a variable name 'SPR'. The matrix has seven columns without labels or text headers. Column 0 of the matrix

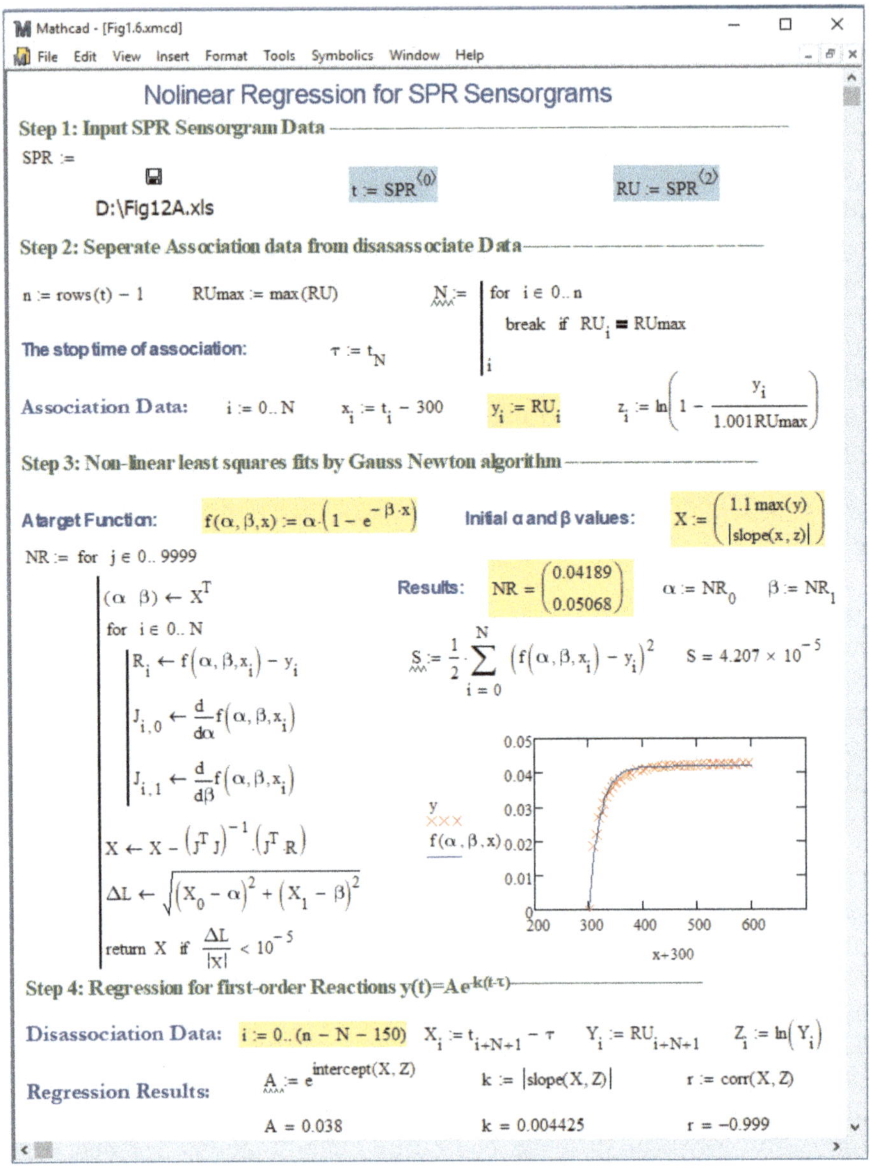

Figure 1.6. A Mathcad program for nonlinear least square fits of sensorgrams by the Gauss–Newton algorithm. The red crosses are half of the experimental data points in an association process, and the blue solid line is the nonlinear regression. Available online at http://iopscience.iop.org/book/978-0-7503-2384-0.

stores time information, and the remaining six columns store the SPR responses to a carbonic anhydrase concentration of 16.6, 8.3, 4.2, 2.1, 1.0 and 0.5 μM respectively. An integer number i in the superscript of '$SPR^{<i>}$' indicates the data in the column i of the SPR matrix. We cannot directly type $SPR^{<i>}$ on the Mathcad interface for coding, but have to use an icon from a matrix toolbar for mathematical operations. The ORIGIN of any matrix in the book is zero (a default value of Mathcad). The data in column 0 of the SPR matrix is then assigned to a vector named 't', and the data in column 2 of the SPR matrix is assigned to a vector named 'RU' for the 8.3 μM curve. By changing the number in the superscript manually from 1 to 6, we can complete nonlinear regressions for all curves in the sensorgram.

In step 2 of the program, 'n:=rows(t) − 1' is used to find the maximum row number of the array t, and 'RUmax:=max(RU)' is used to find the maximum SPR response of a sensorgram curve. Two built-in matrix functions are used. They are rows(B) and max(B) for matrix B. On the same line, a three line sub-program is defined to find N, the row number for the last point of an association curve. Those key words 'for', 'if', and 'break' in the sub-program come from icons in the programming toolbar. The association curve data are then reconstructed with a time shift of 300 s by a 0 to N loop in which 'x_i:= t_i − 300' and 'y_i:= RU$_i$'. Also a matrix z is created from y through a function z_i: $= \ln\left[1 - \frac{y_i}{1.001 \cdot RU_{max}}\right]$.

Step 3 of the program is for a nonlinear regression of the association curve. A target function is defined as $f(\alpha, \beta, x)$: $= \alpha \cdot (1 - e^{-\beta \cdot x})$. Before knowing the optimal values of α and β, how can we estimate them as close as possible to their true values in Gauss–Newton algorithm? A solution comes from matrix z constructed in step 2. The two elements of a matrix X on the same line of the target function are used as the initial guessed values of the two parameters: $\alpha = 1.1\max(y)$ and $\beta = |\mathrm{slope}(x, z)|$. A built-in function slope(A, B) is used to find the slope of vector A and vector B with a linear relationship. The sub-program named 'NR' with two vertical lines has a loop for 9999 iterations. Because of the strategy for choosing the initial parameter values, only 10 iterations are required to get the optimal values: $\alpha = 0.0419°$ and $\beta = 0.0507\ \mathrm{s}^{-1}$. The second line of the sub-program is used to assign the estimate values to α and β respectively. With these initial values, the following loop calculates ($N + 1$) times for the R vector and J matrix based on the formulae shown in equation (1.20). Then a new vector X can be obtained according to the formula (1.19). The distance between the new vector X and the vector (α β) is $\Delta L = \sqrt{(X_0 - \alpha)^2 + (X_1 - \beta)^2}$ and the length of the new vector is $|X|$. If the ratio of ΔL to $|X|$ is greater than or equal to 10^{-5}, the new vector X will be assigned to (α β) again, and the iteration will repeat. When ΔL is less than 10^{-5}, the iteration will be stopped and optimal values of α and β will be returned to the vector NR shown on the right side of the sub-program. With these two optimal parameters, the S value according to equation (1.17) is found to be 4.027×10^{-5}. An x–y plot inside the program shows the original data points (red crosses), and the nonlinear regression results (a blue solid line).

In step 4 of the program, we first pick out the disassociation curve data from the sensorgram curve and shift the time by τ, which is the time to stop the association process and to start an elution procedure. By reconstructing a matrix by $Z = \ln(Y)$, we can use a linear regression in which the last 150 points were not used. As for the sensorgram shown in figure 1.5, the complexes of carbonic anhydrase with its inhibitor do not disassociate completely, the program only used disassociation data for times between 600 s to 750 s. Two built-in functions 'intercept(X, Z)' and 'corr(X, Z)' are used to calculate the linear regression intercept and correlation factor.

Table 1.1. Nonlinear regression results.

$[L_0]$ (μM)	A_∞ (deg)	k_s (s^{-1})	A_τ (deg)	k_d (s^{-1})
16.6	0.0479	0.0697	0.0412	0.004 59
8.3	0.0419	0.0507	0.0384	0.004 43
4.2	0.0354	0.0306	0.0342	0.004 50
2.1	0.0291	0.0195	0.0285	0.004 46
1.0	0.0230	0.0130	0.0220	0.004 31
0.5	0.0195	0.0088	0.0179	0.004 40

Table 1.1 is the nonlinear regression results for the sensorgram shown in figure 1.5. There is a linear relationship between the concentration of carbonic anhydrase and the observable rate constant: $k_s = 0.0038[L_0] + 0.015$. According to the formula (1.14), we get the association rate constant $k_a = 0.0038$ ($\mu M^{-1} \cdot s^{-1}$), and the disassociation rate constant $k_d = 0.015$ (s^{-1}). The affinity of the carbonic anhydrase to its inhibitor immobilized on the gold surface of the sensor chip should be: $K_A = 0.253\ \mu M^{-1}$. However, by choosing the average disassociation rate constant from the six dissociation curves of the sensorgram: $k_d = 0.004\ 45$ (s^{-1}), we have another affinity $K_A = 0.854\ \mu M^{-1}$.

1.2.3 Using double reciprocal plots for affinity measurements

In most cases, elution processes do not follow $A_\tau e^{-k_d(t-\tau)}$ at the elution ends because of non-specific adsorptions and/or other reasons. We could even get different K_A values from the same sensorgram of a ligand–receptor system by choosing different time segments on sensorgram curves. Diffusion processes could delay ligand–receptor interactions. Higher density of receptors on sensing chips might cause side-hindrance to impede binding of ligands to receptors. To obtain perfect sensorgrams, we need to try many times in different conditions. Using double reciprocal plots, we can measure K_A without precisely measuring the association and elution processes. From the formula (1.15), we have an equation for a double reciprocal plot

$$\frac{1}{A_\infty} = \frac{K_D}{\frac{\ell_L + \ell_R}{\ell_R}\Delta Y_R}\frac{1}{[L_0]} + \frac{1}{\frac{\ell_L + \ell_R}{\ell_R}\Delta Y_R} \tag{1.21}$$

In a set of affinity assay experiments with the same kind of sensing chips, we could measure A_∞ values at different ligand concentrations $[L_0]$. According to formula (1.21), we would have a straight line about $1/A_\infty$ to $1/[L_0]$. The ratio of the slope and the intercept of the straight line would be equal to $K_D = 1/K_A$. Figure 1.7 is an example. It is a sensorgram from Lahiri's paper [26] on vancomycin which specifically binds with its receptor Acetyl-Lys-D-Ala-D-Ala immobilized in a dextran matrix on the gold surface of a BIAcore 1000 sensing chip. Gram-positive bacteria synthesize long polymers of *N*-acetylmuramic acid and *N*-acetylglucosamine. These

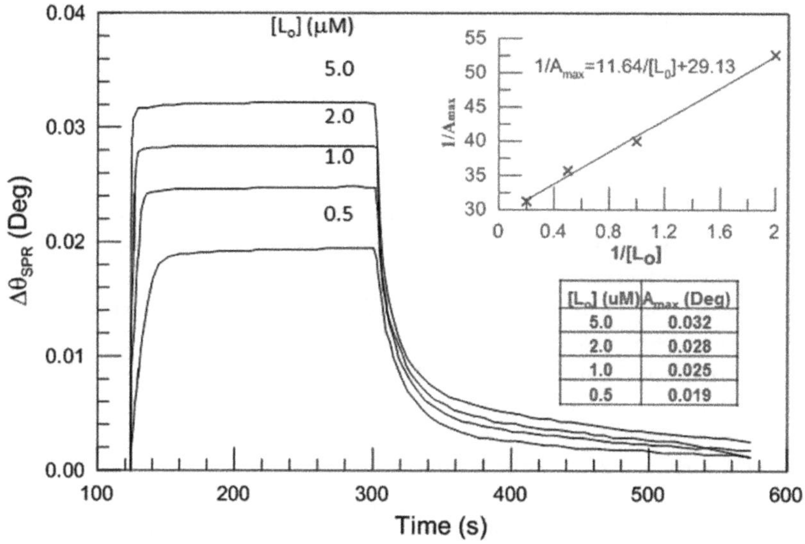

Figure 1.7. A sensorgram in Lahiri's paper [26] on vancomycin molecules (0.5, 1, 2, and 5 μM) react with Acetyl-Lys-D-Ala-D-Ala receptors immobilized in a dextran matrix on a gold surface of a BIAcore 1000 sensing chip. Reprinted with permission from [26]. Copyright (1999) American Chemical Society.

sugar polymer chains must be cross-linked by short peptide chains so that their cell walls are strong enough to resist the osmotic pressure in water. At the end of a short peptide chain, there are two D-alanine residues. Vancomycin is a glyeopeptide antibiotic against Gram-positive bacteria by binding with D-Ala-D-Ala to inhibit cell-wall synthesis. The antibiotic was discovered by Edmund Kornfeld in the soil in Borneo, Indonesia. It is the last resort to fight superbugs. Nonlinear regression results of those curves shown in figure 1.7 were not reliable. Using the maximum SPR response $A_{\max}$ to replace A_{∞}, we find a linear relationship about $1/A_{\max} \sim 1/[L_0]$. The slope and the intercept of the line are 11.64 and 29.13 respectively. From them, we obtain $K_D = 0.4$ μM, and $K_A = 2.5$ μM^{-1}. Sensorgrams are not needed to measure affinities of ligand–receptor interactions.

References

[1] Heidelberger M and Kendall F E 1936 Quantitative studies on antibody purification. I. The dissociation of precipitates formed by pneumococcus specific polysaccharides and homologous antibodies *J. Exp. Med.* **64** 161–72

[2] Miller S L 1953 A production of amino acids under possible primitive earth conditions *Science* **117** 528–9

[3] Watson J D and Crick F H C 1953 Molecular structure of nucleic acids: a structure for deoxyribose nucleic acid *Nature* **171** 737

[4] Witzany G 2008 The viral origins of telomeres and telomerases and their important role in eukaryogenesis and genome maintenance *Biosemiotics* **1** 191–206

[5] Pensa E, Cortés E, Corthey G, Carro P, Vericat C, Fonticelli M H, Benítez G, Rubert A A and Salvarezza R C 2012 The chemistry of the sulfur–gold interface: in search of a unified model *Acc. Chem. Res.* **45** 1183–92

[6] Pauling L, Corey R B and Branson H R 1951 The structure of proteins: Two hydrogen-bonded helical configurations of the polypeptide chain *Proc. Natl. Acad. Sci.* **37** 205–11

[7] Erickson H P 2009 Size and shape of protein molecules at the nanometer level determined by sedimentation, gel filtration, and electron microscopy *Biol. Proc. Online* **11** 32–51

[8] Xiao C and Zeng X 2010 Portable surface plasmon resonance biosensor *Google Patents* (Rochester, MI: Oakland University) 31

[9] Liedberg B, Nylander C and Lundström I 1983 Surface plasmon resonance for gas detection and biosensing *Sens. Actuators A Phys.* **4** 299–304

[10] Rothenhäusler B and Knoll W 1988 Surface plasmon microscopy *Nature* **332** 615–7

[11] Xu Q-M, Wan L-J, Wang C, Bai C-L, Wang Z-Y and Nozawa T 2001 New structure of l-cysteine self-assembled monolayer on Au(111): studies by *in situ* scanning tunneling microscopy *Langmuir* **17** 6203–6

[12] Kuhnle A, Linderoth T R, Schunack M and Besenbacher F 2006 L-Cysteine adsorption structures on Au(111) investigated by scanning tunneling microscopy under ultrahigh vacuum conditions *Langmuir* **22** 2156–60

[13] Xiao C and Luong J H T 2010 A simple mathematical model for electric cell-substrate impedance sensing with extended applications *Biosens. Bioelectron.* **25** 1774–80

[14] Tačić A, Nikolić V, Nikolić L and Savić I 2017 Antimicrobial sulfonamide drugs *Adv. Technol.* **6** 58–71

[15] Fischer E 1894 Einfluss der Configuration auf die Wirkung der Enzyme *Ber. Dtsch. Chem. Ges.* **27** 2985–993

[16] Anfinsen C B 1972 The formation and stabilization of protein structure *Biochem. J.* **128** 737–49

[17] Koshland D E 1995 The key–lock theory and the induced fit theory *Angew. Chem. Int. Ed. Engl.* **33** 2375–8

[18] Steward M W 1984 *Outline Studies in Biology Antibodies: Their Structure and Function* (New York: Chapman and Hall)

[19] Pollard T D 2010 A guide to simple and informative binding assays *Mol. Biol. Cell* **21** 4061–7

[20] Green N M 1963 Avidin. 1. The use of (14-C)biotin for kinetic studies and for assa *Biochem. J.* **89** 585–91

[21] Berg J M, Tymoczko J L and Stryer L 2002 *Biochemistry* 5th edn International Version (San Francisco: Freeman)

[22] Scatchard G 1949 The attraction of proteins for small molecules and ions *Ann. N. Y. Acad. Sci.* **51** 660–72

[23] Klintström S W 1992 *Ellipsometry and Wettability Gradient Surfaces* (Sweden: Linköping University)

[24] Chincholi B S, Havlik A J and Vold R D 1974 Specific refractive index increments of polymer systems a four wavelengths *J. Chem. Eng. Data* **19** 148–52

[25] Feijter J A D, Benjamines J and Veer F A 1978 Ellipsometry as a tool to study the adsorption behavior of synthetic and biopolymer at the air water interface *Biopolymers* **17** 1759–72

[26] Lahiri J, Isaacs L, Tien J and Whitesides G M 1999 A strategy for the generation of surfaces presenting ligands for studies of binding based on an active ester as a common reactive intermediate: a surface plasmon resonance study *Anal. Chem.* **71** 777–90
[27] Kakeya N, Aoki M, Kamada A and Yata N 1969 Biological activities of drugs. VI. Structure-activity relationship of sulfonamide carbonic anhydrase inhibitors. (1) *Chem. Pharm. Bull.* **17** 1010–8
[28] Supuran C T and Scozzafava A 2000 Carbonic anhydrase inhibitors: aromatic sulfonamides and disulfonamides act as efficient tumor growth inhibitors *J. Enzyme Inhib.* **15** 597–610
[29] http://fourier.eng.hmc.edu/e176/lectures/NM/node36.html
[30] https://en.wikipedia.org/wiki/Gauss%E2%80%93Newton_algorithm
[31] https://en.wikipedia.org/wiki/Newton%27s_method

IOP Publishing

Super Optical Biosensors

Caide Xiao

Chapter 2

Eigen-waves in optical systems

Visible light is used as a probe in optical biosensors to detect biological films with a thickness of only a few nanometers. From Maxwell's equations in simple media, we obtain harmonic oscillation functions of monochromatic light waves. The interactions of a light wave with electrons inside a medium determine the refractive index of the medium. For metals, the refractive indices are complex numbers which are useful in surface plasmon resonance. From boundary conditions, we get eigen-waves in optical systems, and find visible light cannot be directly used to excite surface plasmons.

Solid, liquid and gas are three common states of matter. At extreme high temperatures, all atoms of matter are vaporized to become electrons and nuclei. The gases on the surface of the Sun or in a shining neon light tube are plasmas, the fourth state of matter. Metals such as gold, silver, copper and aluminum are good conductors, in which valence electrons freely flow in crystal lattices constructed by positively charged nuclei. The valence electrons shared by all nuclei of a metal behave just like free electrons in hot plasmas. Each of these metals can be treated as a 'cold' plasma. Although each ion in a plasma has electric charge(s), plasmas are neutral at its equilibration condition. However, any disturbance producing net charges at any location inside the plasma will spread to the whole space of the plasma by Coulomb's force. Electrons in the plasma will collectively oscillate. We know that light waves have the quantum name 'photons' given by Compton [1] in 1928. The collective oscillation of electron density in a metal was given a quantum name 'plasmon' by Pines and Bohm [2] in 1952 based on their studies about energy loss of fast electrons passing through metal foils. 'Surface plasmon' is a collective oscillation of electron density on a metal surface predicated by Ritchie [3] in 1957 for experiments of electron energy losses. Interactions between surface plasmons and visible light were first observed by Wood [4] in 1902 in his 'anomalous' reflected light patterns from a diffraction grating mirror. In 1968, Otto [5], Kretschmann and Raether [6, 7] found methods to excite surface plasmons by evanescent light waves

doi:10.1088/2053-2563/ab0823ch2

created by total reflections. In 1983, Liedberg [8] applied Kretschmann's method (configuration) to excite surface plasmons on gold/water interfaces for biosensing applications. This kind of optical biosensor is called a 'surface plasmon resonance biosensor' (SPR biosensor). With a relative higher sensitivity and a simple structure, SPR biosensors are widely used to monitor biomolecular interactions in real time. SPR biosensors can be used as balances to measure less than a nanogram per square millimeter of biomolecules on a gold/water interface. If antigen molecules specifically related to a sickness are immobilized on a gold surface of a SPR sensing chip, we can use a SPR biosensor to investigate if there are specific antibodies in a patient's blood. It is not necessary to label the blood samples with radioactive elements or fluorescent dyes. This is one of many advantages of the SPR biosensing technique. To understand the principle of SPR biosensors and other optical biosensors, we have to start from optical refractive indices of matter.

2.1 Complex refractive indices of noble metals

Light can pass through water and glasses, but not metals, because metals strongly reflect light. In SPR biosensors, gold films with a thickness of 50 nm are used for sensing chips. The behaviors of matter to light are dependent on their optical refractive indices. Isaac Newton in 1666 proved that white light was a mixture of different colored lights. Today we know that each color light is an electromagnetic wave with a frequency or a wavelength. In free space, light at any frequency (monochromatic) propagates in a constant speed of c (=3×10^8 m s^{-1}). The speed of a monochromatic light in a transparent medium with a refractive index n is: $v = c/n$. For example, the refractive index of water to a red light from a He–Ne laser (λ = 632.8 nm) is 1.3317, so the speed of the red light in water is about $v = 2.253 \times 10^8$ m s^{-1}. But the refractive indices of gold and silver to the laser are complex numbers ($0.194\,25 + j3.5912$) and ($0.031\,254 + j4.3885$) respectively. Metals can be used for mirror coatings because of their complex refractivity indices.

2.1.1 Bulk plasmons in metal

In any simple harmonic oscillation system, such as a ball attached to spring, a pendulum, or a LC electric circuit, there is always a restore force to set the system to oscillate around its equilibrium position. The motion of the system can be described by a harmonic function:

$$y(t) = A \cdot \cos(\omega t + \phi_0) \tag{2.1}$$

where A is the amplitude of the harmonic motion, $\omega = 2\pi/T$ is the angular frequency of the system with a unit of [rad s^{-1}], $T = 1/f$ is the period of the motion with a unit of seconds [s], f is the frequency of the motion with a unit of [Hz] or [s^{-1}], ϕ_0 is the initial phase of the motion, and the phase of the motion at time t is $\phi = \omega t + \phi_0$. The amplitude A and initial phase ϕ_0 can be determined by an initial position and an initial speed of the harmonic motion.

It is convenient to use Euler's formula to describe a simple harmonic motion for mathematics calculations, where e is Euler's number (~2.718 28) and $j = \sqrt{-1}$. By choosing the real part of equation (2.2), we get equation (2.1)

$$y(t) = A \cdot e^{j(\omega t + \phi_0)} \tag{2.2}$$

A metal could be treated as a cold plasma with positively charged nuclei in its crystal lattices and negatively charged moving electrons. Any disturbance causing a local net electrical charge will affect all electrons in the plasma by Coulomb force. In a simple case, if a disturbance causes all electrons in the metal to shift a distance z in the direction perpendicular to two parallel surfaces of a metal slab (figure 2.1), the metal's top and bottom surface will have an electrical charge surface density $-qNz$ and qNz respectively, where N is the free electron density of the metal and $-q$ is the electric charge of an electron ($q = 1.6 \times 0^{-19}$ C). According to Gauss's law, an electrical field is then established inside the metal with a strength of $E = qNz/\varepsilon_0$, where ε_0 is the vacuum permittivity with an approximate value of 8.8541×10^{-12} Farads per meter [F m^{-1}]. Any free electron inside the metal will be subjected to a force by the electrical field: $-qE$. The force will pull the electrons back to their equilibrium positions, but with overshoots. The situation is similar to a ball hanging on an end of a spring in a simple harmonic motion system. According to Newton's second law, we could write a motion equation for an electron: $m\frac{d^2z}{d^2t} = -\frac{1}{\varepsilon_0}q^2Nz$. A solution of the equation is $z = z_0e^{-j\omega_p t}$, where z_0 is the amplitude of the electrons collective oscillation, and $\omega_p = \sqrt{\frac{Nq^2}{\varepsilon_0 m}} = 2\pi f_p$ is the angular frequency of the plasma oscillation

$$f_p = \frac{1}{2\pi}\sqrt{\frac{Nq^2}{\varepsilon_0 m}} = 8.973\sqrt{N} \tag{2.3}$$

The plasma oscillation in a 3D space has the quantum name 'bulk plasmon' to distinguish it from 'surface plasmon'. It is a longitudinal wave, just like sound waves traveling in the air. The oscillation frequency f_p is only dependent on the electron density in the plasma, and it is a critical parameter to decide if an electromagnetic wave can go through or be reflected by the bulk plasmon. The atmosphere 60 km above Earth is an ionosphere, in which air molecules are ionized by solar radiations. The electron density in the ionosphere changes with solar activities from 10^9 to

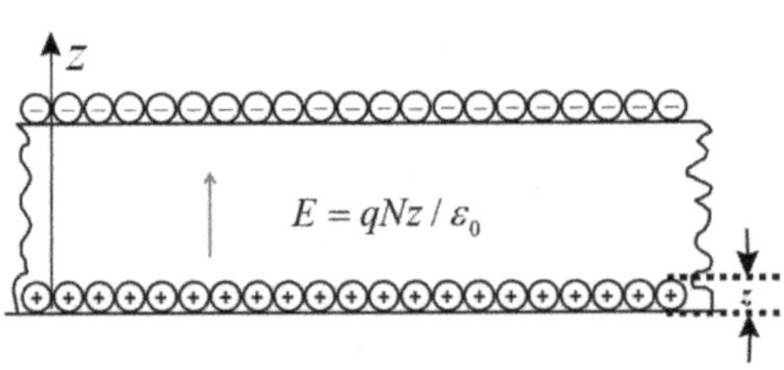

Figure 2.1. A cross section of a metal slab in which electrons are collectively shifted a distance z by a disturbance, so an electric field is established inside the metal by surface charges on the top and bottom of the slab.

10^{12} m^{-3}. By choosing $N = 10^{12}$ m^{-3}, we get $f_p = 9$ MHz. Short wave radio signals below the frequency can be reflected by the ionosphere to reach the opposite side of the Earth. Communication between satellites and their ground stations must use electromagnetic waves of higher frequencies to penetrate through the ionosphere. When an electromagnetic wave's frequency is higher than f_p, even electrons in the plasma cannot move fast enough to respond to the electromagnetic wave, so the plasma is transparent to the electromagnetic wave. Many metals are shiny because their plasmon oscillation frequencies are in ultraviolet frequencies. For example, the density of gold is 19.3 g cm^{-3}, and its standard atomic weight is 197. Since each gold atom provides only one valence electron, we can calculate to get gold's free electron density $N = 5.90 \times 10^{28}$ m^{-3} and its plasmon frequency $f_p = 2.168 \times 10^{15}$ Hz. The wavelength (λ) of the plasma oscillation of gold is about 138 nm, an ultraviolet light. If electrons in gold are forced to oscillate with the frequency of visible light, the light will be reflected.

The energy and momentum of a bulk plasmon are $\hbar\omega_p$ and $\hbar k$ respectively, where $\hbar = h/(2\pi)$ is the reduced Planck constant ($h = 6.626 \times 10^{-34}$ J·s), and $k = 2\pi/\lambda$ is the wavenumber of the plasmon. The energy of a gold plasmon is 1.444×10^{-18} J. The energy unit Joule is too big for photons and plasmons. In the nano-world, the energy unit of electronvolt (1 eV = 1.6×10^{-19} J) is often used. It is the potential energy obtained by an electron passing through an electric field with one volt difference. The energy of a gold plasmon is equal to 9.023 eV. The energy of a photon of a He–Ne laser is 1.963 eV. By using electrons with energy of several keV to bombard metal foil, we can observe plasmon excitations by recording electron energy loss spectroscopy.

2.1.2 Light waves or photons in simple media

The Sun is the major light source on the Earth. Light is an electromagnetic radiation (EMR) with different wavelengths (λ). What our eyes can see is visible light, a very narrow band of EMR. Using a prism, we can see the full spectrum of visible light from blue (λ = 400 nm) to red (λ = 700 nm). Ancient human beings could only produce visible and infrared lights by burning photosynthesis products and fossil fuels. Now we can make devices to produce EMR of different wavelengths: γ-rays, x-rays, ultraviolet light, visible light, infrared light, microwaves, short waves and long waves. Electromagnetic waves were predicted by Maxwell's electromagnetic equations, and were confirmed by Hertz through electrical spark experiments. Whenever electrical particles are accelerated in space, EMR will be created, and EMR carry energy and momentum away at the speed of light in a vacuum.

Laser was a great invention of the 20th century. Its name came from an abbreviation of 'light amplification by stimulated emission of radiation'. A light emitted by a laser is coherent (in phase), monochromatic (single wavelength) and collimated (very narrow beam in space). In a linear polarized laser ray, there are two synchronized oscillations: an electric field E vector and a magnetic flux density B vector. In figure 2.2, the electric field vector is described by a harmonic function $\vec{E} = E_0 \cos(kz - \omega t) \cdot \vec{e}_x$, and the magnetic flux density vector is described by

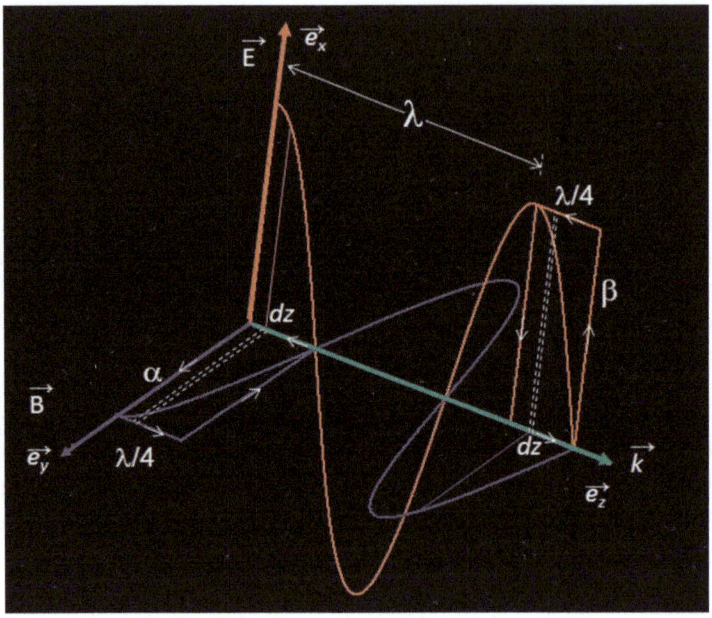

Figure 2.2. A monochromatic light wave traveling along the z direction with its electric field E vector oscillates on the xoz plane, its magnetic flux density B vector oscillates on the yoz plane, and its wavenumber vector ($k = 2\pi/\lambda$) is perpendicular to both E and B vectors.

another harmonic function $\vec{B} = B_0 \cos(kz - \omega t) \cdot \vec{e}_y$, where E_0 and B_0 are amplitudes of the two oscillations respectively. Since the wavenumber vector $\vec{k}$ is parallel with $\vec{e}_z$, the light propagates along the positive z direction of a 3D xyz Cartesian coordinate system, where $\vec{e}_x$, $\vec{e}_y$ and $\vec{e}_z$ are base unit vectors of x, y and z axes respectively. Just like the x, y and z axes, the three vectors $\vec{E}$, $\vec{B}$ and $\vec{k}$ are perpendicular to each other, and their orientations are decided by the right-hand rule. In the figure, there are two virtual rectangles in the 3D xyz space. The length of their short sides along the z-axis is $\lambda/4$. The length of the longer side along the y-axis is α, and the length of the longer side along the x-axis is β. Using the paths marked by arrows on the two rectangles and Maxwell's equations, we can calculate the speed of light.

2.1.2.1 Maxwell's equations in simple media

Maxwell's equations are a group of four physics laws expressed in integral and partial differential equations respectively. The first equation is Gauss's law: electric charges create electric flux which is proportional to the electric charges inside a space with a volume V surrounded by its surface. In the equation (2.4), ρ_V is the volume density of free electric charge with the SI unit of Coulombs per cubic meter [C m^{-3}], and D with an arrow over it represents a vector called electric flux density with the SI unit [C m^{-2}]. This vector is also called the electric displacement field, introduced by Maxwell. Gauss's law in its partial differential form tells us that the divergence of an electric flux density at any point in space is equal to the free electric volume density at that point

$$\oiint \vec{D} \cdot d\vec{s} = \oiiint \rho_v dv \qquad \nabla \cdot \vec{D} = \rho_v \tag{2.4}$$

An upside-down capital delta symbol ∇ is called 'del', and it is a mathematical operator used in vector calculus. In a 3D Cartesian coordinate, a del could be written as: $\nabla = \vec{e}_x \frac{\partial}{\partial x} + \vec{e}_y \frac{\partial}{\partial y} + \vec{e}_z \frac{\partial}{\partial z}$. Using ∇ directly onto any scalar function $f(x, y, z)$, we obtain a vector: the gradient of the scalar function

$$\nabla f(x, y, z) = \frac{\partial f}{\partial x}\vec{e}_x + \frac{\partial f}{\partial y}\vec{e}_y + \frac{\partial f}{\partial z}\vec{e}_z \tag{2.5}$$

A dot production of a del to a vector $\vec{F} = F_x\vec{e}_x + F_y\vec{e}_y + F_z\vec{e}_z$ is a scalar function: the divergence of the vector

$$\nabla \cdot \vec{F} = \frac{\partial}{\partial x}F_x + \frac{\partial}{\partial y}F_y + \frac{\partial}{\partial z}F_z \tag{2.6}$$

A cross production of a del to a vector $\vec{F}$ is called the curl of the vector which is still a vector

$$\nabla \times \vec{F} = \begin{vmatrix} \vec{e}_x & \vec{e}_x & \vec{e}_x \\ \dfrac{\partial}{\partial x} & \dfrac{\partial}{\partial y} & \dfrac{\partial}{\partial z} \\ F_x & F_y & F_z \end{vmatrix} \tag{2.7}$$

The dot product of two dels is a Laplace operator which can act on any scalar or any vector function

$$\nabla^2 = \nabla \cdot \nabla = \frac{\partial^2}{\partial x^2} + \frac{\partial^2}{\partial y^2} + \frac{\partial^2}{\partial z^2} \tag{2.8}$$

The divergence of any vector's curl is equal to zero

$$\nabla \cdot (\nabla \times \vec{F}) = 0 \tag{2.9}$$

The curl of any vector's curl can be decomposed into two parts

$$\nabla \times (\nabla \times \vec{F}) = \nabla(\nabla \cdot \vec{F}) - \nabla^2\vec{F} \tag{2.10}$$

For example, if an electric charge Q is homogenously distributed in a sphere with a radius R, and the center of the sphere is at the origin of a Cartesian coordinate, a static electric field inside and outside of the sphere can be obtained from Gauss's law

$$\vec{D}(r) = \begin{cases} \dfrac{Q}{4\pi}\dfrac{1}{R^3}\vec{r} & (r \leqslant R) \\ \dfrac{Q}{4\pi}\dfrac{1}{r^3}\vec{r} & (r \geqslant R) \end{cases} \tag{2.11}$$

where r with an arrow over it is a vector pointing from the origin to a point (x, y, z) with a length of $r = \sqrt{x^2 + y^2 + z^2}$. Only inside the sphere $\nabla \cdot \vec{D}(r)$ is not equal to zero, but $\rho_v = Q(\frac{4\pi}{3}R^3)^{-1}$, the free electric charge density.

The second equation is Gauss's law of electromagnetism: The total magnetic flux out of a space with a closed surface is equal to zero, or the divergence of any magnetic flux density at any point in space is equal to zero, because there is no magnetic monopole in the Universe. In the equation, B with an arrow over it represents magnetic flux density with the SI unit of Webers per square meter, [Wb m^{-2}] or [Tesla]

$$\oiint \vec{B} \cdot d\vec{s} = 0 \qquad \nabla \cdot \vec{B} = 0 \tag{2.12}$$

The third equation is Faraday's law: A change of magnetic flux inside an area induces an electromotive force (EMF) in a closed loop around the area, and the EMF direction is determined by Lenz's law. In this equation, E with an arrow over it is the electric field vector with the SI unit volts per meter, [V m^{-1}]. Principles of electric motors, generators, and transformers are all based on Faraday's law. The differential form of Faraday's law says that the curl of an induced electric field at any point is equal to the negative change rate of the magnetic flux density vector there

$$\oint \vec{E} \cdot d\vec{\ell} = -\frac{d}{dt}\oiint \vec{B} \cdot d\vec{s} \qquad \nabla \times \vec{E} = -\frac{\partial}{\partial t}\vec{B} \tag{2.13}$$

In an example of an application of Faraday's law, we chose the vertical rectangular area on the xoz plane in figure 2.2. The width and height of the rectangle are $\lambda/4$ and β respectively. According to the right-hand rule and the directions of those arrows marked on the rectangle's sides, the normal of the rectangle area is parallel with the y-axis. On the two short sides of the rectangle, $\vec{E}$ is perpendicular to both of them, so the integrations $\int \vec{E} \cdot d\vec{\ell}$ on the two shorter sides are equal to zero. On the longer side at $z = 5\lambda/4$ at time t, we have $\int \vec{E} \cdot d\vec{\ell} = E_0\beta \sin(\omega t)$. At the same time on the longer side at $z = \lambda$, $\vec{E}$ is in the opposite direction to its path: $\int \vec{E} \cdot d\vec{\ell} = -E_0\beta \cos(\omega t)$. Considering all the above results, we have $\oint \vec{E} \cdot d\vec{\ell} = E_0\beta[\sin(\omega t) - \cos(\omega t)]$.

At the same time, the negative change rate of the magnetic flux across the rectangular area is found to be: $-\frac{d}{dt}\oiint \vec{B} \cdot d\vec{s} = \frac{\omega}{k}\beta B_0[\sin(\omega t) - \cos(\omega t)]$.

If we set $t = 0$ in the above integrations, the calculations are simpler. From these integration results, we find a relationship between the amplitudes of the E and B oscillations shown in the figure, where v is the phase velocity of the electromagnetic wave with a wavelength λ and an angular frequency ω

$$B_0 = \frac{k}{\omega} E_0 = \frac{1}{v} E_0 \tag{2.14}$$

There is a constitutive equation of matter for the relationship between $\vec{D}$ and $\vec{E}$. If a medium is linear, homogenous and isotropic, it is a simple medium in which their relationship is linear

$$\vec{D} = \varepsilon \vec{E} = \varepsilon_r \varepsilon_0 \vec{E} \tag{2.15}$$

where ε is a constant called the permittivity of matter. Furthermore, the constant is defined as: $\varepsilon = \varepsilon_r \varepsilon_0$, where ε_0 is the vacuum permittivity. A dimensionless constant ε_r is the relative permittivity of the medium, and usually is a function changed with light frequency or wavelength. For example, the ε_r of water at microwave frequencies is about 88, but at visible light frequencies, it is only about 1.77. In vacuum and air, ε_r is a constant: $\varepsilon_r = 1$.

The fourth equation is Ampère–Maxwell's law: The integration of a magnetic field along a closed loop is equal to free currents plus a displacement current crossing an area surrounded by the loop. In this equation, H with an arrow over it represents the magnetic field with the SI unit of Ampere's per meter, [A m^{-1}]; J with an arrow over it represents free electric current density with the SI unit Ampere's per square meter, [A m^{-2}]. The differential form of Ampère–Maxwell's law tells us that at any point in space the changing rate of the electric flux density plus a free current density is equal to the curl of the magnetic field

$$\oint \vec{H} \cdot d\vec{\ell} = \oiint \vec{J} \cdot d\vec{s} + \frac{d}{dt} \oiint \vec{D} \cdot d\vec{s} \qquad \nabla \times \vec{H} = \vec{J} + \frac{\partial}{\partial t} \vec{D} \tag{2.16}$$

There is also a constitutive equation of matter for the relationship between $\vec{B}$ and $\vec{H}$. In a simple medium, their relationship is linear

$$\vec{B} = \mu \vec{H} = \mu_r \mu_0 \vec{H} \tag{2.17}$$

In the above equations, $\mu = \mu_r \mu_0$ is a constant called the permeability of matter, where μ_0 is the vacuum permeability with an approximate value of 1.2566×10^{-6} Henrys per meter [H m^{-1}], μ_r is the relative permeability of the medium, a dimensionless constant. In a vacuum and in air, it is a constant: $\mu_r = 1$. For non-magnetic matter, μ_r is also very close to 1.

Now we can use the integration form of Ampère–Maxwell's law to the horizontal rectangular area on the *yoz* plan in figure 2.2. The width and length of the rectangle are $\lambda/4$ and α respectively. Because the two short paths with a length of $\lambda/4$ are perpendicular to $\vec{B}$, we have $\int \vec{H} \cdot d\vec{\ell} = 0$. At time $t = 0$ and on the longer path with a length α at $z = \lambda/4$, we also get $\int \vec{H} \cdot d\vec{\ell} = 0$ because $B = 0$. The integration of $\vec{H}$

around the four sides of the rectangle at $t = 0$ only contributes to the integration along the longer path at $z = 0$: $\oint \vec{H} \cdot d\vec{\ell} = B_0\alpha/\mu$. The first integration part of the right side of the formula (2.16) is equal to zero in case there is no free current in the simple medium, $\vec{J} = 0$. The second part on the right side of equation (2.16) at $t = 0$ is the electric flux crossing the rectangular area surrounded by the paths: $\frac{d}{dt}\oiint \vec{D} \cdot d\vec{s} = \alpha\varepsilon\frac{d}{dt}\int_0^{\frac{\lambda}{4}} E_0 \cos(kz - \omega t)dz\,|_{t=0} = \alpha\varepsilon v E_0$.

From the integration results of both sides of equation (2.16), we also have another relationship between the amplitudes of the two oscillations:

$$B_0 = \varepsilon\mu v E_0 \tag{2.18}$$

Combining equations (2.14) and (2.18), we find the phase speed of light in a simple medium

$$\mathrm{v} = \frac{1}{\sqrt{\varepsilon\mu}} = \frac{c}{\sqrt{\varepsilon_r\mu_r}} = \frac{c}{n} \tag{2.19}$$

where $n = \sqrt{\varepsilon_r\mu_r}$ is the refractive index of a simple medium, and is a function of light frequency or wavelength. In free space and air, the two constants ε and μ are ε_0 and μ_0 respectively, so a monochromatic light wave's exact phase speed c is 299 792 458 m s^{-1} in free space.

2.1.2.2 Monochromatic light waves in simple media

Now we will manipulate Maxwell's equations in their differential forms to get light wave equations, and prove their solutions are really harmonic functions (cosine or sine waves). Using $\nabla\cdot$ to act on both side of equation (2.16), we get:

$$\nabla \cdot (\nabla \times \vec{H}) = \nabla \cdot \vec{J} + \frac{\partial}{\partial t}\nabla \cdot \vec{D} \tag{2.20}$$

The left side is equal to zero because the divergence of any vector's curl is zero. With Gauss's law, the right side of the above equation is equal to $\nabla \cdot \vec{J} + \frac{\partial}{\partial t}\rho_{\mathrm{v}}$. Now we obtain the electric continuity equation: the divergence of a free electric current density at any point in space is equal to the losing rate of the free electric charge volume density at that point

$$\nabla \cdot \vec{J} = -\frac{\partial}{\partial t}\rho_{\mathrm{v}} \tag{2.21}$$

By using $\nabla \times$ to both side of Faraday's law (2.13), we have:

$$\nabla \times (\nabla \times \vec{E}) = -\frac{\partial}{\partial t}\nabla \times \vec{B} \tag{2.22}$$

The left side of this equation can be separated into two parts: $\nabla(\nabla \cdot \vec{E}) - \nabla^2\vec{E}$. According to Gauss's law (2.4), $\nabla \cdot \vec{E}$ is equal to zero for $\rho_{\mathrm{v}} = 0$ in a source free medium. In the source free medium, there is no free electric charge, so there is no free electric current: $\vec{J} = 0$. From equations (2.16), (2.17) and (2.19), $-\frac{\partial}{\partial t}\nabla \times \vec{B}$ is changed to: $-\frac{n^2}{c^2}\frac{\partial^2}{\partial^2 t}\vec{E}$. Then we have a light wave equation about the electric field that varies in space with time

$$\nabla^2\vec{E} - \left(\frac{n}{c}\right)^2 \frac{\partial^2}{\partial t^2}\vec{E} = 0 \tag{2.23}$$

A solution of the wave equation is described by equation (2.24), where $\vec{E}_0$ is a constant vector depending on light wave intensity and direction

$$\vec{E}(\vec{r}, t) = \vec{E}_0 e^{j(\vec{k}\cdot\vec{r}-\omega t)} \tag{2.24}$$

In the same way, we can also find a wave equation and its solution about the magnetic flux density $\vec{B}(\vec{r}, t)$. Then from Faraday's law (2.13), the magnetic flux density can be expressed by the electric field

$$\vec{B}(\vec{r}, t) = \frac{n}{c}\frac{\vec{k}}{k} \times \vec{E}_0 e^{j(\vec{k}\cdot\vec{r}-\omega t)} \tag{2.25}$$

where $\vec{k} = k_x\vec{e}_x + k_y\vec{e}_y + k_z\vec{e}_z$ is the light wavenumber vector with a length of $2\pi/\lambda$, and $\vec{k} \cdot \vec{r} = k_x x + k_y y + k_z z$. The electric field and magnetic flux density are harmonic waves with the same phase $\phi = \vec{k} \cdot \vec{r} - \omega t$. The light wave is a plane wave, because at any time all points in space with the same phase are on a plane in a 3D Cartesian coordinate. In the simple case shown in figure 2.2, the light wave propagates in the z-axis direction with $k_x = 0$, $k_y = 0$ and $k_z = k$, so the light wave can be described by two simple harmonic functions:

$$\vec{E}(\vec{r}, t) = \mathrm{Re}[E_0 e^{j(\vec{k}\cdot\vec{r}-\omega t)}]\vec{e}_x = E_0 \cos(kz - \omega t)\vec{e}_x \tag{2.26}$$

$$\vec{B}(\vec{r}, t) = \mathrm{Re}[B_0 e^{j(\vec{k}\cdot\vec{r}-\omega t)}]\vec{e}_y = \frac{n}{c}E_0 \cos(kz - \omega t)\vec{e}_y \tag{2.27}$$

Both the electric field and the magnetic flux density of the light are transverse waves. A plane constructed by $\vec{E}$ and $\vec{k}$ is the polarization plane of the light. A light with only one polarized plane is a linear polarized light. A combination of $\vec{E}_x(\vec{r}, t) = E_0 \cos(kz - \omega t)\vec{e}_x$ and $\vec{E}_y(\vec{r}, t) = E_0 \cos(kz - \omega t + \phi_0)\vec{e}_y$ is still a solution of the wave equation (2.23), where ϕ_0 is the phase difference between $\vec{E}_x$ and $\vec{E}_y$. When ϕ_0 is equal to an integer times π, we have a linear polarized light, and when ϕ_0 is equal to any other values, we have an elliptical polarized light. Using a polarizer, we can convert an un-polarized light into a linear polarized light, but with a sacrifice of light intensity. From the magnitude of the Poynting vector, we can calculate light intensity: the time average of light energy flux. The energy density [J m^{-3}] stored in an electric field and a magnetic field is $\frac{\varepsilon}{2}E^2$ and $\frac{1}{2\mu}B^2$ respectively. From

$|\vec{B}| = \sqrt{\varepsilon\mu}|\vec{E}|$, we have $\frac{1}{2\mu}B^2 = \frac{\varepsilon}{2}E^2$, so the energy density of the light is equal to εE^2. The energy flux of the light is equal to $\varepsilon v E^2 = \sqrt{\frac{\varepsilon}{\mu}}E^2$, and its time average is equal to:

$$\langle \vec{S} \rangle = \frac{1}{2}\,\mathrm{Re}\,\{\vec{E} \times \vec{H}^*\} = \frac{1}{2}\sqrt{\frac{\varepsilon}{\mu}}E_0^2\frac{\vec{k}}{k} \tag{2.28}$$

where $\vec{H}^*$ is a complex conjugate of the magnetic field $\vec{H}$, and E_0 is the amplitude of the electric field of the light. For example, if a He–Ne laser tube emits a 5 mW light beam with a radius of 2 mm, its average energy flux is 398 W m^{-2}, its amplitude of electric field (E_0) is 548 V m^{-1}, and its amplitude of magnetic flux density (B_0) is 1.83×10^{-6} Wb m^{-2} or Tesla.

A perfect harmonic light wave propagates in space forever without a beginning and an end. This wave does not exist because all light waves or photons must be emitted by sources and adsorbed by media somewhere, sometime. The start and end of light waves will produce light waves with different frequencies. Even two light waves with the same amplitude but very small differences in their wavelength ($\Delta\lambda$) or frequency ($\Delta\omega$), their superposition is a modulated wave. Combining these two harmonic waves: $y_1 = Ae^{j[(k+\frac{\Delta k}{2})z-(\omega+\frac{\Delta\omega}{2})t]}$ and $y_2 = Ae^{j[(k-\frac{\Delta k}{2})z-(\omega-\frac{\Delta\omega}{2})t]}$, we get:

$$y = y_1 + y_2 = A\cos\left(\frac{\Delta k}{2}z - \frac{\Delta\omega}{2}t\right)e^{j(kz-\omega t)} \tag{2.29}$$

The amplitude of the combining wave is slowly changed with time in space. The phase speed of the amplitude is a group velocity of the combining light wave to distinguish it from phase velocity $v_p = \omega/k$

$$v_g = \frac{\Delta\omega}{\Delta k} \tag{2.30}$$

A relationship between ω and k is a dispersion relation of the medium. According to equation (2.19), the dispersion relation of a simple medium is: $\omega = kc/n$, and the group velocity of a light wave in the medium is:

$$v_g = \frac{c}{n} - \frac{ck}{n^2}\frac{\Delta n}{\Delta k} \tag{2.31}$$

In vacuum and in air, $n = 1$ is a constant, so the phase speed and group speed are the same. If a medium's refractive index changes with light wavelength, the medium is a dispersive medium. Water, glass and metals are all dispersive media. That is why we can see rainbows.

2.1.3 Forced harmonic oscillations and complex refractive indices

Chromatic dispersion is an optical phenomenon caused by the refractive index of matter changing with light frequency or wavelength. We can use a complex dielectric function $\varepsilon_r(\omega) = \varepsilon_r'(\omega) + j\varepsilon_r''(\omega)$ to describe dispersive properties of matter, where $\varepsilon_r'(\omega)$ and $\varepsilon_r''(\omega)$ are, respectively, the real part and imaginary parts of the dielectric constant at a frequency ω. From $n = \sqrt{\varepsilon_r}$ and $n = \eta + j\kappa$, we explicitly have their relationships:

$$\varepsilon_r' = \eta^2 - \kappa^2 \tag{2.33}$$

$$\varepsilon_r'' = 2\eta\kappa \tag{2.34}$$

$$\eta = \frac{1}{\sqrt{2}}\left[\sqrt{(\varepsilon_r')^2 + (\varepsilon_r'')^2} + \varepsilon_r'\right]^{1/2} \tag{2.35}$$

$$\kappa = \frac{1}{\sqrt{2}}\left[\sqrt{(\varepsilon_r')^2 + (\varepsilon_r'')^2} - \varepsilon_r'\right]^{1/2} \tag{2.36}$$

From equation (2.24), we can see that the electric field amplitude of a light will be attenuated if it is passing through a medium with a complex refractive index. For simplification, we just assume that the light travels in the z direction and the electric field is in the xoz plane shown in figure 2.2

$$\vec{E}(\vec{r}, t) = \vec{E}_0 e^{-\kappa\frac{\omega}{c}z} e^{j\left(\eta\frac{\omega}{c}z - \omega t\right)} \tag{2.37}$$

The depth of the light penetrating into a metal is called the 'skin depth' at which the electric field amplitude is attenuated to $1/e$ of its surface value (or light intensity is reduced to $1/e^2 \approx 13\%$)

$$\delta = \frac{c}{\kappa\omega} \tag{2.38}$$

2.1.3.1 Lorentz model of dielectric refractive indices

When visible light is propagating in matter, electrons and nuclei of atoms of matter will respond to the electrical field and magnetic flux density of the light. Compared to an electron, a nucleus is so heavy that its movement can be ignored. A force subjected to a moving electron in matter by the light is a Lorentz force

$$\vec{F} = -q(\vec{E} + \vec{v} \times \vec{B}) \tag{2.39}$$

In the above formula, $\vec{v}$ is the speed of the moving electron, and q is the absolute electric charge of the electron. According to the formula (2.14), the second part of the Lorentz force contributed by the magnetic flux density can be ignored if the speed of the moving electron is far less than the speed of light. It is only necessary to consider the effects of the electrical field of the light in non-magnetic matter.

Newton's second law can be used to derive a relationship about refractive indices of matter and light frequencies. The movement of an electron on an outside shell of an atom at the origin of a Cartesian coordinate could be:

$$m\frac{d^2\vec{r}}{d^2t} = -q\vec{E} - m\omega_0^2\,\vec{r} - m\xi\frac{d\vec{r}}{dt} \tag{2.40}$$

where $\vec{r}$ is the shift of the electron relative to its positive nucleus, m is the electron mass ($m = 9.1 \times 10^{-31}$ kg). In the case $|\vec{r}| \ll \lambda$, the electric field of the light could be treated as a homogeneous field at the origin of the Cartesian coordinate system: $\vec{E} = \vec{E}_o e^{-j\omega t}$, so that the first part on the right side of the above equation $-q\vec{E}$ is the force contributed by the light to the electron. The second part, $-m\omega_0^2\vec{r}$ is a quasi-elastic force contributed by the nucleus to the electron in a resonant oscillation with a natural frequency ω_0. The third part, $-m\xi\frac{d\vec{r}}{dt}$ is a damping force subjected to the moving electron by collisions (ξ: damping constant). An explicit solution of the equation (2.40) is:

$$\vec{r} = \left(\frac{-q}{\varepsilon_0 m}\right)\frac{1}{\omega_0^2 - \omega^2 - j\xi\omega}\varepsilon_0\vec{E}_o e^{-j\omega t} \tag{2.41}$$

Then we can get an induced electric dipole moment $-q\vec{r}$ contributed by the moving electron, and the polarization density: $\vec{P} = -Nq\vec{r}$, where N is the electron density in the medium. The polarization density $\vec{P}$ in a simple medium is proportional to and aligned with the external electric field:

$$\vec{P} = \chi_e\varepsilon_0\vec{E} \tag{2.42}$$

where χ_e is the electric susceptibility of the medium. From the above two formulae, we can explicitly write χ_e as:

$$\chi_e = \left(\frac{Nq^2}{\varepsilon_o m}\right)\frac{1}{\omega_0^2 - \omega^2 - j\xi\omega} \tag{2.43}$$

By taking $\vec{D} = \varepsilon_r\varepsilon_0\vec{E}$ and $\vec{P} = \chi_e\varepsilon_0\vec{E}$ into a formula about electric flux density: $\vec{D} = \varepsilon_o\vec{E} + \vec{P}$, we obtain a formula for the relative permittivity of the simple media $\varepsilon_r = 1 + \chi_e$, where $\omega_p = \sqrt{\frac{Nq^2}{\varepsilon_0 m}}$ is the plasma oscillation angular frequency

$$\varepsilon_r = 1 + \frac{\omega_p^2}{\left(\omega_0^2 - \omega^2\right) - j\xi\omega} \tag{2.44}$$

For transparent media, such as air, water and glass, light adsorption can be ignored, so the damping force could be set to be zero ($\xi = 0$). Formula (2.44) can be simplified for transparent media as:

$$\varepsilon_r = 1 + \frac{\omega_p^2}{\left(\omega_0^2 - \omega^2\right)} \tag{2.45}$$

Usually a molecule of a transparent medium has more than one natural frequency ω_0 in near-infra-red, visible, and ultraviolet spectral regions (10^{14} – 10^{15} Hz). With

$n = \sqrt{\varepsilon_r}$ and in a condition $\omega < \omega_0$ for a visible light, we can see that the higher the frequency, the bigger the refractive index is. This is a normal dispersion function. From a database on the website: http://refractiveindex.info/, we can find refractive indices of many kinds of optical media, such as glasses, plastics, liquids and metals. For example, to He–Ne laser (632.8 nm), the refractive index of a heavy flint glass ZF7, crown glass BK7 and water is 1.7988, 1.5151 and 1.3317 respectively.

2.1.3.2 Drude model of metal refractive indices

Since valence electrons in metal are not bound to nuclei of atoms, the quasi-elastic force is equal to zero, so their natural frequency (ω_0) is equal to zero. The formula (2.44) can be simplified to a Drude model for metal relative permittivity

$$\varepsilon_r = \left(1 - \frac{\omega_p^2}{\omega^2 + \xi^2}\right) + j\left(\frac{\xi}{\omega}\right)\frac{\omega_p^2}{\omega^2 + \xi^2} \tag{2.46}$$

For any monochromatic light in a visible wavelength, the angular frequency ω is bigger than 2.7×10^{15} rad s^{-1}. The damping force is caused by internal collisions of electrons in the metal, and the damping constant [9, 10] ξ is less than 10^{14} s^{-1}. In fact, we can derive a formula about ξ from metal conductivity in direct current (DC) conditions: σ_{DC}. Using the formula (2.41), we have the electron's speed when the frequency (ω) of the electric field is set to be zero

$$\frac{d\vec{r}}{dt} = \frac{q}{\varepsilon_0 m}\frac{-j\omega}{\omega^2 + j\xi\omega}\varepsilon_0\vec{E}_0 e^{-j\omega t}\,\Big|_{\omega=0} = \frac{-q}{m\xi}\vec{E} \tag{2.47}$$

Since a DC current flows in a direction opposed to the electron moving direction, a minus symbol is added in front of N (electron density) to get a DC current density inside the metal

$$\vec{J} = -Nq\frac{d\vec{r}}{dt} = \frac{\varepsilon_0}{\xi}\frac{Nq^2}{\varepsilon_0 m}\vec{E} \tag{2.48}$$

According to Ohm's law, the electric current density has a linear relationship with the electric field

$$\vec{J} = \sigma_{DC}\vec{E} \tag{2.49}$$

From the above two formulae, we get an explicit formula to calculate the damping constant

$$\xi = \frac{\varepsilon_0\omega_p^2}{\sigma_{DC}} \tag{2.50}$$

With gold's DC conductivity $\sigma_{DC} = 4.1 \times 10^7$ [S m^{-1}] and $\omega_p = 1.369 \times 10^{16}$ [rad s^{-1}], we find the electron damping constant of gold: $\xi = 4.045 \times 10^{13}$ s^{-1}. The wavelength of an electromagnetic wave at this frequency is about 7.4 μm (mid-infrared light).

2.1.3.3 $\omega \ll \xi$ (microwaves)

For an electromagnetic wave, such as a microwave whose frequency is far less than the damping constant, formula (2.46) can be simplified as:

$$\varepsilon_r = \varepsilon_r' + j\varepsilon_r'' = \left(1 - \frac{\sigma_{DC}^2}{\varepsilon_0^2\omega_p^2}\right) + j\frac{\sigma_{DC}}{\omega\varepsilon_0} \tag{2.51}$$

It is easy to find $\varepsilon_r'' \gg |\varepsilon_r'|$, so ε_r' can be ignored. For example, to a microwave at 1 GHz, the relative permittivity of gold is $\varepsilon_r = -1.144 \times 10^5 + j7.37 \times 10^7$ according to formula (2.51). From formulae (2.35) and (2.36), the refractive index of metal for low frequency electromagnetic waves can be obtained with $\eta \approx \kappa = (\frac{\varepsilon_r''}{2})^{\frac{1}{2}} = (\frac{\sigma_{DC}}{2\omega\varepsilon_0})^{\frac{1}{2}}$. With formula (2.38), we have the skin depth of a low frequency electromagnetic wave penetrating into a metal with DC conductivity σ_{DC}

$$\delta = \sqrt{\frac{2}{\mu_0\omega\sigma_{DC}}} \tag{2.52}$$

For example, the skin depth of a 10 GHz microwave to gold is about 0.786 μm. The reflectivity of a light directly shining with an incident angle $\theta = 0°$ on a metal is $R = |n - 1|^2/|n + 1|^2$. With $\eta^2 \approx \kappa^2 \gg 1$, the reflectivity can be explicitly written as the Hagen–Rubens relation [11]

$$R = \left|\frac{\eta + j\kappa - 1}{\eta + j\kappa + 1}\right|^2 = 1 - \frac{4\eta}{(\eta + 1)^2 + \kappa^2} \simeq 1 - \frac{2}{\eta} = 1 - \sqrt{\frac{8\varepsilon_0\omega}{\sigma_{DC}}} \tag{2.53}$$

Metals with higher conductivity reflect more energy back for EMR beyond the far-infra-red region ($\lambda > 10$ μm) [12]. For a perfect metal, σ_{DC} is infinite, the skin depth should be zero and the reflectivity is equal to one. That is why the James Webb space telescope, the successor of the Hubble space telescope, uses gold mirrors for infrared images.

2.1.3.4 $\omega > \omega_p$ (UV light to x-rays and γ-rays)

In another extreme condition, the electromagnetic wave could be UV light, or even x-rays and γ-rays. The imaginary part of ε_r in equation (2.46) can be ignored, so the metal dielectric constant becomes a real positive number whose value is less than one. For example, alkali metals are transparent to UV light. For $\omega \gg \omega_p$, ε_r is close to one. A metal to light with such extreme high frequency is just like vacuum. That is why x-rays and γ-rays can penetrate through metals

$$\varepsilon_r = 1 - \frac{\omega_p^2}{\omega^2} \tag{2.54}$$

By taking the above formula into $k^2 = \varepsilon_r\omega^2/c^2$ we get the light dispersion function in metal:

$$\omega^2 = \omega_p^2 + c^2k^2 \tag{2.55}$$

With this formula, it is easy to see that the cut-off frequency of light traveling in metal is ω_p. Under the cut-off frequency, k is a complex number with only an imaginary part. Above the cut-off frequency, light travels in metal with a group speed less than light speed in free space: $v_g = \frac{d\omega}{dk} = c/\sqrt{1 + (\frac{\omega_p}{ck})^2}$.

2.1.3.5 $\xi < \omega < \omega_p$ (infra-red light and visible red light)

In the condition, ε_r' is negative number whose absolute value is bigger than 10.0, and ε_r'' is a positive number whose value is around 1.0. Light cannot propagate through metal, but will be reflected on the metal surface. For example, the ε_r of gold [13] and silver [14] to He–Ne laser light (632.8 nm) are $(-11.8550 + j1.6771)$ and $(-18.630 + j0.442\,48)$ respectively

$$\varepsilon_r = \left(1 - \frac{\omega_p^2}{\omega^2}\right) + j\left(\frac{\xi}{\omega}\right)\frac{\omega_p^2}{\omega^2} \tag{2.56}$$

The Drude model for noble metals' refractive indices is only valid for red light and infrared light [15, 16]. Photons with higher frequencies than red light can also interact with inner shell electrons inside the atoms of noble metals. Because of the contribution of these bound electrons to atom polarizations, the Drude model about metal refractive indices does not match experimental data for light with higher frequencies.

2.2 Boundary conditions and surface plasmons

On a smooth interface between two different media, a linear polarized light will be reflected and refracted. Because of the boundary conditions of electromagnetic waves, the intensities of the reflected light and refracted light are dependent on the orientation of the polarization plane of the light. In figure 2.3, an xoy plane of a 3D Cartesian coordinate is coincident with the interface between two semi-infinite media which are non-magnetic materials. Their refractive indices are $n_1 = \sqrt{\varepsilon_{1r}}$ and $n_2 = \sqrt{\varepsilon_{2r}}$ respectively. The z-axis of the Cartesian system points from medium 1 to medium 2. A normal vector $\vec{n}$ of the interface is parallel with the z-axis, and they are in the same direction. If the light incident angle is θ_1, then according to Snell's law, the reflection angle is equal to the incident angle, and the refraction angle is equal to $\theta_2 = \arcsin[\frac{n_1}{n_2}\sin(\theta_1)]$.

The incident plane of the light is a plane with the normal vector $\vec{n}$ and the light wavenumber vector $\vec{k}$. Referencing to the incident plane, we have the incident light in a transverse magnetic (TM) mode or in a transverse electric (TE) mode. For an incident light in the TM mode, its magnetic field $\vec{H}$ is perpendicular to the incident plane, and for an incident light in the TE mode, its electric field $\vec{E}$ is perpendicular to the incident plane. From Maxwell's equations, we have four boundary conditions:

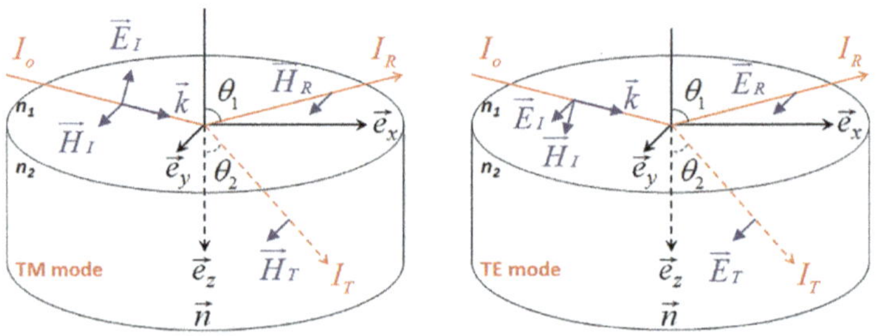

Figure 2.3. A linear polarized light shining on a smooth interface between two semi-infinite media could be in a transverse magnetic mode (TM mode, left) or in a transverse electric mode (TE mode, right).

$$\begin{cases} \vec{n} \cdot (\vec{D}_2 - \vec{D}_1) = \sigma_s \\ \vec{n} \cdot (\vec{B}_2 - \vec{B}_1) = 0 \\ \vec{n} \times (\vec{E}_2 - \vec{E}_1) = 0 \\ \vec{n} \times (\vec{H}_2 - \vec{H}_1) = \vec{J}_s \end{cases} \tag{2.57}$$

Coming from Gauss's law (2.4), the first boundary condition indicates that the difference between the normal components of $\vec{D}$ on two sides of an interface is equal to the free surface electric charge density σ_s [C m^{-2}] on the interface. If there is no free electric charge on the interface, the normal components of $\vec{D}$ across the interface are continuous: $\varepsilon_{1r}E_{1z} = \varepsilon_{2r}E_{2z}$. The second boundary condition is about magnetic flux density. The normal components of $\vec{B}$ across the interface are continuous: $B_{1z} = B_{2z}$.

The third boundary condition is derived from Faraday's law (2.13). It tells us that the tangent components of the electric fields on two sides of the interface are continuous. For example, the TE mode light waves shown on the right of figure 2.3 have $\vec{E}_1 = (E_I + E_R)\vec{e}_y$ and $\vec{E}_2 = E_T\vec{e}_y$, so we have $E_I + E_R = E_T$ at the interface $z = 0$, where E_I, E_R, E_T are electric fields of the incident light, the reflected light and the refractive light respectively. If medium 2 in figure 2.3 is a perfect metal, the electric field inside the metal should be equal to zero: $E_T = 0$. In a simple condition with $\theta_1 = 0$, there is no difference between the TM and TE mode waves. The incident light is: $\vec{E}_I = E_0e^{j(kz-\omega t)}\vec{e}_y$. To fulfill the boundary condition on the interface, we should have $\vec{E}_R = -E_0e^{j(-kz-\omega t)}\vec{e}_y$, and get a stationary light wave in medium 1: $\vec{E}_I + \vec{E}_R = 2E_0 \sin(kz)\sin(\omega t)\vec{e}_y$. Using an 88 MHz microwave ($\lambda = 3.4$ m), a former MIT Professor, Walter Lewin, vividly demonstrated a stationary electromagnetic wave in a YouTube video lecture [17]. Along the $-z$ direction, there are nodes and antinodes, where the microwave intensities were minimum and maximum respectively, and the distance between a node and its neighbor antinode is $\lambda/4$.

The last boundary condition tells us that the difference of the tangent components of the magnetic fields ($\vec{H}$) on side 2 and side 1 of an interface is equal to the linear free electric current density $\vec{J}_s$ with a unit of [A m^{-1}]. If there is no free electric current on the interface for a TM mode light shown on the left of figure 2.3, we have

$H_I + H_R = H_T$ at the interface, where H_I, H_R, H_T are magnetic fields of the incident light, the reflected light and the refractive light respectively.

2.2.1 Eigen-waves of a system with two semi-infinite simple media

In the system with two semi-infinite simple media shown in figure 2.3, there is no free electric charge and no free electric current. Eigen-waves are possible solutions of the wave equation (2.23) in the system

$$\vec{E}(\vec{r}, t) = \begin{cases} \vec{E}_1^{\,0} e^{j(k_s x + \alpha z - \omega t)} & (z \leqslant 0) \\ \vec{E}_2^{\,0} e^{j(k_s x + \beta z - \omega t)} & (z \geqslant 0) \end{cases} \tag{2.58}$$

In this solution, the wavenumber vectors in medium 1 and 2 are $\vec{k}_1 = k_s\vec{e}_x + \alpha\vec{e}_z$, $\vec{k}_2 = k_s\vec{e}_x + \beta\vec{e}_z$, and the electric field vectors are $\vec{E}_1^{\,0} = E_{1x}^0\vec{e}_x + E_{1y}^0\vec{e}_y + E_{1z}^0\vec{e}_z$, $\vec{E}_2^{\,0} = E_{2x}^0\vec{e}_x + E_{2y}^0\vec{e}_y + E_{2z}^0\vec{e}_z$ respectively, and they can be obtained from the four boundary conditions in equation (2.57) through the following mathematical manipulations. By taking the possible solutions into the wave equation $\nabla^2\vec{E} - \left(\frac{n}{c}\right)^2 \frac{\partial^2}{\partial t^2}\vec{E} = 0$, we have the following two equations about α, β and k_s

$$\alpha^2 + k_s^2 = \varepsilon_{1r}\frac{\omega^2}{c^2} \tag{2.59}$$

$$\beta^2 + k_s^2 = \varepsilon_{2r}\frac{\omega^2}{c^2} \tag{2.60}$$

Without free electric charge in the system, we have $\nabla \cdot \vec{E} = 0$ in both medium according to Gauss's law, so we obtain $E_{1z}^0 = -\frac{k_s}{\alpha}E_{1x}^0$ and $E_{2z}^0 = -\frac{k_s}{\beta}E_{2x}^0$. We can also get $\vec{B}$ in both media from Faraday's law: $\vec{B} = -\frac{j}{\omega}\nabla \times \vec{E}$

$$\vec{E}(\vec{r}, t) = \begin{cases} \left(E_{1x}^0,\ E_{1y}^0,\ \dfrac{-k_s}{\alpha}E_{1x}^0\right) e^{j(k_s x + \alpha z - \omega t)} & (z \leqslant 0) \\ \left(E_{2x}^0,\ E_{2y}^0,\ \dfrac{-k_s}{\beta}E_{2x}^0\right) e^{j(k_s x + \beta z - \omega t)} & (z \geqslant 0) \end{cases} \tag{2.61}$$

$$\vec{B}(\vec{r}, t) = \begin{cases} \left(-\dfrac{\alpha}{\omega}E_{1y}^0,\ \dfrac{1}{\omega}\dfrac{\varepsilon_{1r}}{\alpha}\dfrac{\omega^2}{c^2}E_{1x}^0,\ \dfrac{k_s}{\omega}E_{1y}^0\right) e^{j(k_s x + \alpha z - \omega t)} & (z \leqslant 0) \\ \left(-\dfrac{\beta}{\omega}E_{2y}^0,\ \dfrac{1}{\omega}\dfrac{\varepsilon_{2r}}{\beta}\dfrac{\omega^2}{c^2}E_{2x}^0,\ \dfrac{k_s}{\omega}E_{2y}^0\right) e^{j(k_s x + \beta z - \omega t)} & (z \geqslant 0) \end{cases} \tag{2.62}$$

According to the boundary condition about the continuous normal components of the magnetic flux densities: $\vec{n} \cdot (\vec{B}_2 - \vec{B}_1) = 0$, we have $B_{1z} = B_{2z}$ on the interface of

$z = 0$ because the normal vector of the interface is in the same direction of the z-axis shown in figure 2.3

$$\frac{k_s}{\omega}E_{1y}^0 = \frac{k_s}{\omega}E_{2y}^0 \tag{2.63}$$

Without free current on the interface $z = 0$, we simplify the boundary condition about the magnetic fields to $\vec{n} \times (\vec{H}_2 - \vec{H}_1) = 0$. According to the formula (2.62), the boundary condition became $B_{1x} = B_{2x}$ because $\vec{B} = \mu_0\vec{H}$ in simple media

$$\frac{-\alpha}{\omega}E_{1y}^0 = \frac{-\beta}{\omega}E_{2y}^0 \tag{2.64}$$

Solving equations (2.63) and (2.64), we get $E_{1y}^0 = E_{2y}^0 = 0$. Since the tangent components of electric fields are continuous on the interface: $\vec{n} \times (\vec{E}_2 - \vec{E}_1) = 0$, we also have the following equation

$$E_{1x}^0 = E_{2x}^0 \tag{2.65}$$

Without free electric charge in the system, we know $\vec{n} \cdot (\vec{D}_2 - \vec{D}_1) = 0$ on the interface. Explicitly it is: $D_{2z} = D_{1z}$. With $\vec{D} = \varepsilon_0\varepsilon_r\vec{E}$ for simple media, the boundary condition can be rewritten explicitly as $\varepsilon_{1r}E_{1z}^0 = \varepsilon_{2r}E_{2z}^0$. According to the formula (2.61), we know $E_{1z}^0 = \frac{-k_s}{\alpha}E_{1x}^0$ and $E_{2z}^0 = \frac{-k_s}{\beta}E_{2x}^0$, so we find the third relationship of α and β.

$$\frac{\varepsilon_{1r}}{\alpha} = \frac{\varepsilon_{2r}}{\beta} \tag{2.66}$$

Combining equations (2.59), (2.60) and (2.66) into a group, we solve it to get the wavenumber vectors' components in the two media

$$\alpha = \frac{\omega}{c}\sqrt{\frac{\varepsilon_{1r}^2}{\varepsilon_{1r} + \varepsilon_{2r}}} \tag{2.67}$$

$$\beta = \frac{\omega}{c}\sqrt{\frac{\varepsilon_{2r}^2}{\varepsilon_{1r} + \varepsilon_{2r}}} \tag{2.68}$$

$$k_s = \frac{\omega}{c}\sqrt{\frac{\varepsilon_{1r}\varepsilon_{2r}}{\varepsilon_{1r} + \varepsilon_{2r}}} \tag{2.69}$$

Finally, we have the electric fields and magnetic flux densities in the system with two semi-infinite simple media. It is in TM mode, not in TE mode

$$\vec{E}(\vec{r},t) = \begin{cases} \left(1,\ 0,\ -\sqrt{\dfrac{\varepsilon_{2r}}{\varepsilon_{1r}}}\right) E_{1x}^{0} e^{j(k_s x + \alpha\, z - \omega t)} & (z \leqslant 0) \\ \left(1,\ 0,\ -\sqrt{\dfrac{\varepsilon_{1r}}{\varepsilon_{2r}}}\right) E_{1x}^{0} e^{j(k_s x + \beta\, z - \omega t)} & (z \geqslant 0) \end{cases} \tag{2.70}$$

$$\vec{B}(\vec{r},t) = \begin{cases} (0,\ 1,\ 0)\dfrac{\sqrt{\varepsilon_{1r}+\varepsilon_{2r}}}{c} E_{1x}^{0} e^{j(k_s x + \alpha\, z - \omega t)} & (z \leqslant 0) \\ (0,\ 1,\ 0)\dfrac{\sqrt{\varepsilon_{1r}+\varepsilon_{2r}}}{c} E_{1x}^{0} e^{j(k_s x + \beta\, z - \omega t)} & (z \geqslant 0) \end{cases} \tag{2.71}$$

(1) Eigen-waves of a system with two transparent media

In a system with two semi-infinite transparent media, α, β and k_s are all real numbers. The ratio between k_s and α is n_2/n_1, from which we get the incident angle of a linear polarized light in TM mode in medium 1: $\theta_b = \arctan(\frac{n_2}{n_1})$, the Brewster angle. The refractive angle θ_2 in medium 2 ($z > 0$) shown in figure 2.3(left) is decided by the ratio between k_s and β: $\theta_2 = \arctan(\frac{n_1}{n_2})$, which can also be derived from Snell's law with the incident angle θ_b. From equations (2.67) to (2.69), it is easy to prove: $\theta_b + \theta_2 = 90°$. For an air/water system to a red light ($\lambda = 632.8$ nm), we have $\alpha = \frac{3.773}{\lambda}$, $\beta = \frac{6.691}{\lambda}$, $k_s = \frac{5.024}{\lambda}$, $\theta_b = 53.1°$ and $\theta_2 = 36.9°$ by using $n_1 = 1$ and $n_2 = 1.3317$. The eigen-waves of the system are two linear polarized light waves in TM mode as shown below. The parameter E_{1x}^{0} is a constant decided by the light intensity

$$\vec{E}(\vec{r},t) = \begin{cases} \left(1,\ 0,\ \dfrac{1.3317}{-1}\right) E_{1x}^{0} e^{j(5.024\frac{x}{\lambda} + 3.773\frac{z}{\lambda} - \omega t)} & (z \leqslant 0) \\ \left(1,\ 0,\ \dfrac{-1}{1.3317}\right) E_{1x}^{0} e^{j(5.024\frac{x}{\lambda} + 6.691\frac{z}{\lambda} - \omega t)} & (z \geqslant 0) \end{cases}$$

$$\vec{B}(\vec{r},t) = \begin{cases} \left(0,\ \dfrac{1.665}{c},\ 0\right) E_{1x}^{0} e^{j(5.024\frac{x}{\lambda} + 3.773\frac{z}{\lambda} - \omega t)} & (z \leqslant 0) \\ \left(0,\ \dfrac{1.665}{c},\ 0\right) E_{1x}^{0} e^{j(5.024\frac{x}{\lambda} + 6.691\frac{z}{\lambda} - \omega t)} & (z \geqslant 0) \end{cases}$$

(2) Eigen-waves of a system with a transparent medium and a metal

If medium 1 is a transparent material and medium 2 is a noble metal such as gold or silver, the eigen-waves of the system only exist in a thin layer close to the interface of the two media, because α, β and k are all complex numbers. The imaginary parts of these complex numbers cause the light intensity to decay exponentially with distance. For example, for an air/gold system to a red light at $\lambda = 670$ nm, $\varepsilon_{1r} = 1.0$ and $\varepsilon_{2r} = -13.4 + j1.4$, we have $\alpha = \frac{1}{\lambda}(0.100 - j1.776)$, $\beta = \frac{1}{\lambda}(1.147 + j23.936)$, and $k_s = \frac{1}{\lambda}(6.529 + j0.027)$. From these parameters, it is easy to write the eigen-waves of the system explicitly

$$\vec{E}(\vec{r}, t) = \begin{cases} (1, 0, 3.671e^{-j1.623})E_{1x}^{0}e^{-0.027\frac{x}{\lambda}+1.776\frac{z}{\lambda}+j(6.529\frac{x}{\lambda}+0.1\frac{z}{\lambda}-\omega t)} & (z \leqslant 0) \\ (1, 0, 0.272e^{j1.623})E_{1x}^{0}e^{-0.027\frac{x}{\lambda}-23.936\frac{z}{\lambda}+j(6.529\frac{x}{\lambda}+1.147\frac{z}{\lambda}-\omega t)} & (z \geqslant 0) \end{cases} \quad (2.72)$$

$$\vec{B}(\vec{r}, t) = \begin{cases} \left(0, \dfrac{3.533}{c}, 0\right)E_{1x}^{0}e^{-0.027\frac{x}{\lambda}+1.776\frac{z}{\lambda}+j(6.529\frac{x}{\lambda}+0.1000\frac{z}{\lambda}+1.515-\omega t)} & (z \leqslant 0) \\ \left(0, \dfrac{3.533}{c}, 0\right)E_{1x}^{0}e^{-0.027\frac{x}{\lambda}-23.936\frac{z}{\lambda}+j(6.529\frac{x}{\lambda}+1.147\frac{z}{\lambda}+1.515-\omega t)} & (z \geqslant 0) \end{cases} \quad (2.73)$$

By defining the decay length as the distance at which electromagnetic field amplitude decreases to 37.8% ($= e^{-1}$), we find the decay length on the interface at the x-axis direction is 37λ (24 810 nm). At directions perpendicular to the interface in the air and in the gold, the decay lengths are 0.563λ (377 nm) and 0.042λ (28 nm) respectively. By comparing the decay lengths along the x-axis and z-axis directions, we can see the light decays in the z-axis are more significant, especially in the metal.

In the air, E_{1z} and E_{1x} have a phase difference -0.517π ($= -1.623$). At any location close to the interface, the electric field $\vec{E}_1$ in the air rotates in the anticlockwise direction with time. Since the phase difference between E_{2z} and E_{2x} is 0.517π, $\vec{E}_2$ at any location inside the gold rotates in a clockwise direction with time. An animation about rotating electric fields of eigen-waves on a dielectric/metal interface can be found in Wikipedia [18]. The electric fields of the eigen-waves in the system have both transversal and longitudinal components. But their magnetic fields only have transversal components in the y-axis direction with a phase difference to E_{1x}: 0.482π ($= 1.515$).

The electric fields of the eigen-waves can induce polarized electric charges on the metal surface. The polarized electric charge density on the interface is:

$$\sigma_P = -\vec{n} \cdot (\vec{P}_2 - \vec{P}_1)|_{z=0} = \varepsilon_0[(\varepsilon_{1r} - 1)E_{1z} - (\varepsilon_{2r} - 1)E_{2z}]\,|_{z=0} \quad (2.74)$$

For the above air/gold system at $\lambda = 670$ nm, the surface electric charge density on the gold surface can be obtained by choosing the real part of equation (2.74):

$$\sigma_P = 3.48 \times 10^{-11}E_{1x}^{0}e^{-0.027\frac{x}{\lambda}}\cos\left(6.529\frac{x}{\lambda} - \omega t + 0.486\pi\right) \quad (2.75)$$

Since gold is a good conductor, the fluctuation of electron density propagates on the gold surface as a longitudinal wave. Surface electron's collective oscillations were first predicted by Ritchie in his paper about energy losses of fast electrons going through thin metal films [3]. The existence of the phenomenon was experimentally demonstrated two years later by Powell and Swan [19]. The electron collective oscillation wave on a metal surface has the quantum name ‘surface plasmon’ given by Stern and Ferrell [20]. An electromagnetic wave associated with the surface plasmon is called a ‘surface plasmon wave’.

2.2.2 The dispersion relation of surface plasmon

For a metal with $\varepsilon'_{2r} < 0$ and $|\varepsilon'_{2r}| \gg \varepsilon''_{2r}$, we can simplify the formula (2.69) to:

$$k_s = \frac{\omega}{c}\sqrt{\frac{\varepsilon_{1r}\varepsilon'_{2r}}{\varepsilon_{1r}+\varepsilon'_{2r}}} + j\frac{\omega}{c}\frac{\varepsilon''_{2r}}{2(\varepsilon'_{2r})^2}\left(\frac{\varepsilon_{1r}\varepsilon'_{2r}}{\varepsilon_{1r}+\varepsilon'_{2r}}\right)^{3/2} \tag{2.76}$$

By using the real part of formula (2.56) to replace ε'_{2r} and ignoring the imagery part of equation (2.76), we find the surface plasmon dispersion relation which is normalized by $k_p = \omega_p/c$

$$\frac{k_s}{k_p} = \frac{\omega}{\omega_p}\sqrt{\frac{\varepsilon_{1r}\left(1-\frac{\omega^2}{\omega_p^2}\right)}{1-(1+\varepsilon_{1r})\frac{\omega^2}{\omega_p^2}}} \tag{2.77}$$

If the denominator inside the square root of equation (2.77) is less than zero, k_s will be an imaginary number, and the surface plasmon wave will not exist. The surface plasmon wave has a cut-off frequency at:

$$\omega_{sp} = \frac{\omega_p}{\sqrt{1+\varepsilon_{1r}}} \tag{2.78}$$

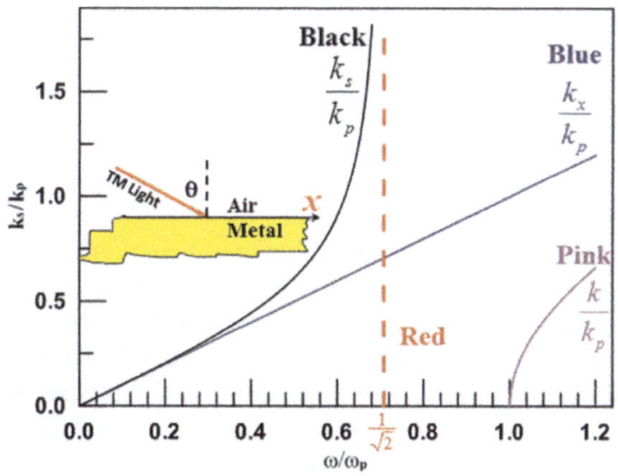

Figure 2.4. Normalized dispersion curves for a surface plasmon wave on an air/metal interface (solid black curve), a light wave traveling in metal (solid pink curve, equation (2.55)), and a TM monochromatic light shining on an air/metal interface (solid blue curve, $\theta = 90°$, $\varepsilon_{1r} = 1$). The vertical dash red line is for the cut-off frequency of surface plasmons.

In the case when medium 1 is air, the cut-off frequency is: $\omega_{sp} = \omega_p/\sqrt{2}$, and the wavelength of the electromagnetic wave at this frequency is about 100 nm. Figure 2.4 shows three normalized dispersion curves: the solid black curve is for the surface plasmon wave on an air/metal interface according to the formula (2.77), the blue straight line is for a light shining on the interface with an incident angle $\theta = 90°$, and a solid pink curve is about a normalized dispersion relation of light traveling in metal according to formula (2.55). The momentum of the surface plasmon is equal to $\hbar k_s$, and in the same direction of the surface plasmon, the photon momentum is equal to $\hbar \frac{\omega}{c} \sin(\theta)$. Since the black curve is above the blue straight line, photons cannot be used to directly excite surface plasmons because the momentum of surface plasmons is 'heavier' than that of photons ($\hbar k_s > \hbar \frac{\omega}{c}$).

References

[1] https://en.wikipedia.org/wiki/Photon
[2] Pines D and Bohm D 1952 A collective description of electron interactions: II. Collective vs individual particle aspects of the interactions *Phys. Rev.* **85** 338
[3] Ritchie R H 1957 Plasma losses by fast electrons in thin films *Phys. Rev.* **106** 874–81
[4] Wood R W 1902 On a remarkable case of uneven distribution of light in a diffraction grating spectrum *Philos. Mag.* **4** 396–402
[5] Otto A 1968 A new method for exciting non-radioactive surface plasma oscillations *Phys. Stat. Sol.* B **26** K99–K101
[6] Kretschmann E J and Reather H 1968 Radiative decay of nonradiative surface plasmon excited by light *Z. Naturforsch.* **23** 2135–6
[7] Raether H 1988 *Surface Plasmons on Smooth and Rough Surfaces and on Gratings* (Berlin: Springer)
[8] Liedberg B, Nylander C and Lundström I 1983 Surface plasmon resonance for gas detection and biosensing *Sens. Actuators A Phys.* **4** 299–304
[9] Kawata S 2001 *Near-field Optics and Surface Plasmon Polaritons* (Heidelberg: Springer)
[10] Novotny L and Hecht B 2006 *Principles of Nano-Optics* 1st edn (New York: Cambridge University Press)
[11] Hagen E and Rubens H 1903 Über die Beziehung des Reflex ions-und Emissionsvermöge ns der Metalle zu ihrem elektrischen Leitvermögen *Ann. Phys. Lpz.* **4** 873–901
[12] Silveira F E M 2011 Metallic absorptivity at normal incidence above far-infrared *Proc. of Progress in Electromagnetics Research Symp.* (*Marrakesh, Morocco*) pp 1029–32
[13] Hagemann H J, Gudat W and Kunz C 1975 Optical constants from the far infrared to the x-ray region: Mg, Al, Cu, Ag, Au, Bi, C, and Al $_2O_3$ *J. Opt. Soc. Am.* **65** 742–4
[14] McPeak K M, Jayanti S V, Kress S J P, Meyer S, Iotti S, Rossinelli A and Norris D J 2015 Plasmonic films can easily be better: Rules and recipes *ACS Photon.* **2** 326–33
[15] Johnson P B and Christy R W 1972 Optical constants of noble metals *Phys. Rev.* **B 6** 4370–9
[16] Maier S A 2007 *Plasmonics: Fundamentals and Applications* (Berlin: Springer)
[17] https://youtube.com/watch?v=dYXjlqwdy44
[18] https://en.wikipedia.org/wiki/Surface_plasmon_polariton
[19] Powell C J and Swan J B 1959 Origin of the characteristic electron energy losses in aluminum *Phys. Rev.* **115** 869–75
[20] Stern E A and Ferrell R A 1960 Surface plasma oscillations of a degenerate electron gas *Phys. Rev.* **120** 130–6

IOP Publishing

Super Optical Biosensors

Caide Xiao

Chapter 3

Surface plasmon resonance biosensors

With reference to a light incident plane, we have a p-polarized light (TM) wave and an s-polarized (TE) wave. Fresnel coefficients and light reflectivity formulae of the two polarized light waves were obtained from boundary conditions. Mathcad programs demonstrated that only TM waves could be used for surface plasmon resonance on a metal surface in Otto configuration or Kretschman configuration. With Rouard's recursive method to numerically simulate light reflectivity in a multilayered system, we found benchmarks of SPR biosensors, and got linear relationships about the shift of the SPR angle to the surface concentration of biological films.

From a transparent dielectric medium, a TM light shining on a smooth gold surface cannot excite eigen-waves, because the photon's momentum is not heavy enough. In 1968, Otto [1] invented a method to tune a photon's momentum by using a prism for internal total reflections, so he could excite surface plasmons with visible light. His smart method is very simple. He set his prism above a gold surface with an air gap. The refractive index of the glass prism is higher ($n_0 > 1.50$). A monochromatic photon traveling inside the prism has a slower speed (c/n_0), so its momentum is increased to $\hbar k = n_0 \hbar \frac{\omega}{c}$. When the light is refracted out of the prism on the glass/air interface with an incident angle θ, the momentum of a photon of the refracted light along the interface is $n_0 \hbar \frac{\omega}{c} \sin(\theta)$ which is possibly bigger than the surface plasmon momentum. During an angle scan process with $\theta > \theta_c$, the photon momentum of an evanescent wave could be heavy enough to match the momentum of a surface plasmon on the air/gold interface. Incident photon energy will then be adsorbed by surface plasmons. On a curve about a relationship between the incident angle and the reflectivity of light on the glass/air interface, there will be an adsorption valley. This optical 'invisible' phenomenon is called surface plasmon resonance (SPR), and the incident angle with a minimum reflectivity is defined as the 'SPR resonance angle' or θ_{SPR}. Because an evanescent wave created by a total internal reflection decays exponentially with a distance perpendicular to the

doi:10.1088/2053-2563/ab0823ch3

interface, the air gap between the prism and the gold surface should be less than one wavelength for surface plasmon resonance.

3.1 Monochromatic light reflection and refraction on an interface

Since a linear polarized light's polarization plan can be parallel with or perpendicular to its incident plane, we have *p*-polarized or *s*-polarized light respectively. A *p*-polarized light has an alternative name, i.e. transverse magnetic (TM) wave, and an *s*-polarized light also has another name, transverse electric (TE) wave. Given an incident monochromatic light's intensity (I_o), angular frequency (ω), incident angle (θ_1) and polarization plane, we can obtain light reflectivity and refractivity with four boundary conditions (2.57) derived from Maxwell's equations.

3.1.1 Transverse magnetic (TM) light

In the 3D-coordinate system shown in figure 3.1, the incident plane of the TM light is on the *xoz* plane. The *y*-axis is not shown in the figure because it is perpendicular to the *xoz* plane and points to readers from the page. The symbol '⊙' indicates that the magnetic field of the incident light at a given time is parallel with the *y*-axis in the same direction

$$\vec{H}_I(\vec{r},\, t) = (0,\, 1,\, 0)H_I^0 e^{j(kx+\alpha z-\omega t)} \tag{3.1}$$

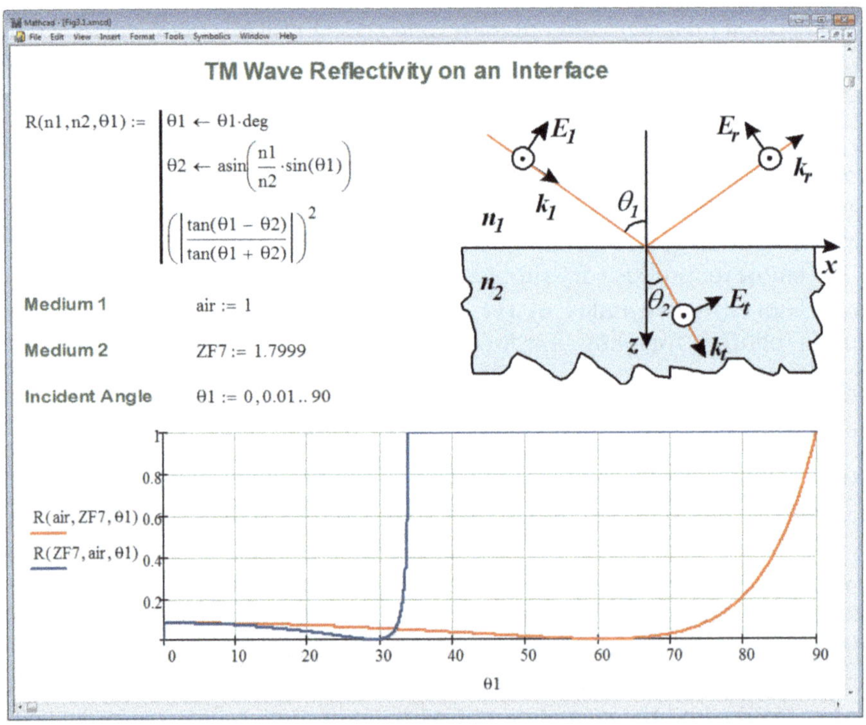

Figure 3.1. A Mathcad program for numerical simulations about a TM wave reflectivity on an air/glass interface (the red curve) and a glass/air interface (the blue curve) for incident angles from 0° to 90°. Available online at http://iopscience.iop.org/book/978-0-7503-2384-0.

In the above expression, the TM wave's magnetic field amplitude H_I^0 is dependent on the intensity of the incident light with a wavelength λ in the air. The wave number vector of the incident light is $\vec{k}_1 = k\vec{e}_x + 0\vec{e}_y + \alpha\vec{e}_z$, where $k = n_1\frac{2\pi}{\lambda}\sin(\theta_1)$, $\alpha = n_1\frac{2\pi}{\lambda}\cos(\theta_1)$. Parameters H_I^0, k and α are known constants decided by the incident light. An explicit expression of the magnetic fields in the system is given below

$$\vec{H}(\vec{r}, t) = \begin{cases} (0, 1, 0)H_I^0 e^{j(kx+\alpha z-\omega t)} + (0, 1, 0)H_R^0 e^{j(kx-\alpha z-\omega t)} & (z \leqslant 0) \\ (0, 1, 0)H_T^0 e^{j(kx+\beta z-\omega t)} & (z > 0) \end{cases} \tag{3.2}$$

where H_R^0 and H_T^0 (the magnetic field amplitudes of the reflected light and the refracted light respectively) are constants that need to be determined. In a space without free electric charge and current, we have a boundary condition on the interface of $z = 0$: $\vec{n} \times (\vec{H}_2 - \vec{H}_1) = 0$, from which we have the first relationship among these magnetic amplitudes

$$H_I^0 + H_R^0 = H_T^0 \tag{3.3}$$

The electric flux density of the light wave can be obtained from a Maxwell equation in a space without a free electric current: $\nabla \times \vec{H} = \frac{\partial}{\partial t}\vec{D}$. By taking equation (3.2) into the left side to calculate $\nabla \times \vec{H}$ and changing the right side from $\frac{\partial}{\partial t}\vec{D}$ to $-j\omega\vec{D}$, we get the following solution

$$\vec{D}(\vec{r}, t) = \begin{cases} \left(\frac{\alpha}{\omega}, 0, \frac{-k}{\omega}\right)H_I^0 e^{j(kx+\alpha z-\omega t)} + \left(\frac{-\alpha}{\omega}, 0, \frac{-k}{\omega}\right)H_R^0 e^{j(kx+\alpha z-\omega t)} & (z \leqslant 0) \\ \left(\frac{\beta}{\omega}, 0, \frac{-k}{\omega}\right)H_T^0 e^{j(kx+\beta z-\omega t)} & (z > 0) \end{cases} \tag{3.4}$$

From a boundary condition $\vec{n} \times (\vec{E}_2 - \vec{E}_1) = 0$ on the interface ($z = 0$), we have an equation about the amplitudes of electric fields: $E_{2x}^0 = E_{1x}^0$, or $\frac{1}{\varepsilon_{1r}}D_{1x}^0 = \frac{1}{\varepsilon_{2r}}D_{2x}^0$. From equation (3.4) and with definitions: $\alpha' = \frac{\alpha}{\varepsilon_{1r}}$ and $\beta' = \frac{\beta}{\varepsilon_{2r}}$, the second relationship among these magnetic amplitudes is:

$$\beta' H_T^0 = \alpha' H_I^0 - \alpha' H_R^0 \tag{3.5}$$

Now we can combine equations (3.3) and (3.5) into a group to find its solutions: $H_R^0 = \frac{\alpha' - \beta'}{\alpha' + \beta'}H_I^0$ and $H_T^0 = \frac{2\alpha'}{\alpha' + \beta\cdot}H_I^0$. From the formulae (2.14) and (2.17), we have the equation about a relationship between the amplitude of $\vec{H}$ and $\vec{E}$ in a simple medium: $H = \sqrt{\frac{\varepsilon}{\mu}}E$. Using the above information, we get Fresnel coefficients about amplitudes of reflected and transmitted light waves at the interface for the TM wave

$$r_p = \frac{E_R^0}{E_I^0} = \frac{H_R^0}{H_I^0} = \frac{\alpha' - \beta'}{\alpha' + \beta'} = \frac{\dfrac{\cos(\theta_1)}{n_1} - \dfrac{\cos(\theta_2)}{n_2}}{\dfrac{\cos(\theta_1)}{n_1} + \dfrac{\cos(\theta_2)}{n_2}} = \frac{\tan(\theta_1 - \theta_2)}{\tan(\theta_1 + \theta_2)} \tag{3.6}$$

$$t_p = \frac{E_T^0}{E_I^0} = \sqrt{\frac{\varepsilon_{1r}}{\varepsilon_{2r}}}\frac{H_T^0}{H_I^0} = \frac{n_1}{n_2}\frac{2\alpha'}{\alpha' + \beta'} = \frac{2\dfrac{\cos(\theta_1)}{n_2}}{\dfrac{\cos(\theta_1)}{n_1} + \dfrac{\cos(\theta_2)}{n_2}} \tag{3.7}$$

The following are explicit expressions of the electromagnetic waves in the system.

$$\vec{H}(\vec{r}, t) = \begin{cases} (0, 1, 0)H_I^0 e^{j(kx+\alpha z-\omega t)} + (0, 1, 0)r_p H_I^0 e^{j(kx-\alpha z-\omega t)} & (z \leqslant 0) \\ (0, 1, 0)\dfrac{n_2}{n_1} t_p H_I^0 e^{j(kx+\beta z-\omega t)} & (z \geqslant 0) \end{cases}$$

$$\vec{E}(\vec{r}, t) = \begin{cases} \left(\dfrac{\alpha}{\omega}, 0, \dfrac{-k}{\omega}\right)\dfrac{H_I^0}{\varepsilon_0\varepsilon_{1r}} e^{j(kx+\alpha z-\omega t)} + \left(\dfrac{-\alpha}{\omega}, 0, \dfrac{-k}{\omega}\right) r_p \dfrac{H_I^0}{\varepsilon_0\varepsilon_{1r}} e^{j(kx-\alpha z-\omega t)} & (z \leqslant 0) \\ \left(\dfrac{\beta}{\omega}, 0, \dfrac{-k}{\omega}\right)\dfrac{n_2}{n_1} t_p \dfrac{H_I^0}{\varepsilon_0\varepsilon_{2r}} e^{j(kx+\beta z-\omega t)} & (z \geqslant 0) \end{cases}$$

To calculate the energy reflectivity R_p and transmission T_p, we have to use the formula (2.28) for average Poynting vectors' components perpendicular to the interface: $\langle S_z \rangle = \frac{1}{2}\sqrt{\frac{\varepsilon}{\mu}} E_0^2 \cos(\theta)$

$$R_p = \frac{n_1 \cos(\theta_1)}{n_1 \cos(\theta_1)} \left| \frac{E_R^0}{E_I^0} \right|^2 = \left| \frac{n_2 \cos(\theta_1) - n_1 \cos(\theta_2)}{n_2 \cos(\theta_1) + n_1 \cos(\theta_2)} \right|^2 \tag{3.8}$$

$$T_p = \frac{n_2 \cos(\theta_2)}{n_1 \cos(\theta_1)} \left| \frac{E_T^0}{E_I^0} \right|^2 = \left| \frac{2\sqrt{n_1 n_2 \cos(\theta_1)\cos(\theta_2)}}{n_2 \cos(\theta_1) + n_1 \cos(\theta_2)} \right|^2 \tag{3.9}$$

With the above two formulae, it is easy to see $R_p + T_p = 1$. But in the case of total reflection, equation (3.9) for T_p is not valid.

A numeric simulation program about a TM wave's reflectivity on an air/glass and a glass/air interface is shown in figure 3.1. The first line of the program is a text region to explain to readers the purpose of the program. On the left side of the second line is a sub-program for calculations of the reflectivity according to the formula (3.8). There are three local variables inside the function: $n1$, $n2$ and $\theta 1$. They are refractive indices of two media and the incident angle of the TM wave inside medium 1 ($0° \leqslant \theta 1 \leqslant 90°$). On the right side of the function, there are three codes aligned to a vertical line. The first code in the function '$\theta 1 \leftarrow \theta 1 \cdot$deg' is used to convert the unit of $\theta 1$ from degrees to radians. An internal constant of Mathcad called 'deg' is equal to $\pi/180$. The second code in the function is used to calculate the refractive angle $\theta 2$ according to Snell's law. The third code of the function returns a reflectivity value when the function is called. Since the value of r_p could be a negative number or a complex number, the absolute value or the amplitude value of r_p is used to calculate $R(n1, n2, \theta 1)$. Below the function are refractive indices of medium 1 and 2 respectively for a He–Ne laser ($\lambda = 632.8$ nm). In line 5 of the program, '$\theta 1 := 0, 0.01 \ldots 90$' is used to set the incident angle changing from 0° to 90° with a step of 0.01°. The red curve in the x–y plot of the program shows the relationship about $R(n1, n2, \theta 1) \sim \theta 1$. For $0° \leqslant \theta 1 < 75°$, the reflectivity on the air/glass interface is less than 10%, and for $\theta 1 \geqslant 75°$, the reflectivity of the system increases exponentially. On the x–y plot we can see that the reflectivity is zero when $\theta 1$ is equal to 60.944°, the Brewster angle.

(1) Brewster angle for eigen-wave excitation

From equation (3.6), we have $r_p = 0$ if $\alpha' = \beta'$ or $n_2 \cos(\theta_1) = n_1 \cos(\theta_2)$. Combining with Snell's law: $n_1 \sin(\theta_1) = n_2 \sin(\theta_2)$, we get $\sin(2\theta_1) = \sin(2\theta_2)$ which has a reasonable solution $\theta_2 = 90° - \theta_1$ and $\theta_1 = \arctan(\frac{n2}{n1})$, the Brewster angle θ_b. With θ_b as incident angle, a TM light wave has zero reflectivity. The electromagnetic waves in the system are simplified by using $E_I^0 = \frac{\alpha}{\varepsilon_0 \varepsilon_{1r} \omega} H_I^0$

$$\vec{E}(\vec{r}, t) = \begin{cases} \left(1, 0, -\sqrt{\dfrac{\varepsilon_{2r}}{\varepsilon_{1r}}}\right) E_{1x}^0 e^{j(kx+\alpha z-\omega t)} & (z \leqslant 0) \\ \left(1, 0, -\sqrt{\dfrac{\varepsilon_{1r}}{\varepsilon_{2r}}}\right) E_{1x}^0 e^{j(kx+\beta z-\omega t)} & (z \geqslant 0) \end{cases}$$

$$\vec{B}(\vec{r}, t) = \begin{cases} (0, 1, 0) \dfrac{\sqrt{\varepsilon_{1r} + \varepsilon_{2r}}}{c} E_{1x}^0 e^{j(kx+\alpha z-\omega t)} & (z \leqslant 0) \\ (0, 1, 0) \dfrac{\sqrt{\varepsilon_{1r} + \varepsilon_{2r}}}{c} E_{1x}^0 e^{j(kx+\beta z-\omega t)} & (z \geqslant 0) \end{cases}$$

These are two synchronized oscillations of an eigen-wave in a system with two simple media as shown in equations (2.70)–(2.71), where $k = n_1 \frac{2\pi}{\lambda} \sin(\theta_b)$, $\alpha = n_1 \frac{2\pi}{\lambda} \cos(\theta_b)$ and $\beta = n_2 \frac{2\pi}{\lambda} \sin(\theta_b)$. The eigen-waves of a system with two simple media can be excited by a TM light with θ_b as its incident angle.

(2) Critical angle for total reflection

The blue curve in figure 3.1 is for reflectivity of the TM light wave in a system with ZF7 glass as medium 1 and air as medium 2. In this system the Brewster angle θ_b is equal to 29.056°, and the refractive angle θ_2 is bigger than the incident angle θ_1. When the refractive angle θ_2 is equal to 90°, the incident angle θ_1 is called the critical angle θ_c for total reflection of the TM wave. On the blue curve in figure 3.1, the critical angle is equal to 33.751°

$$\theta_c = \arcsin\left(\frac{n_2}{n_1}\right) \tag{3.10}$$

In the case where $\theta_1 > \theta_c$, the TM wave is totally reflected back into medium 1. With $n_1 \sin(\theta_1) > n_2$ and $(n_1\frac{2\pi}{\lambda})^2 \sin^2(\theta_1) + \beta^2 = (n_2\frac{2\pi}{\lambda})^2$, β becomes a complex number: $\beta = j\frac{2\pi}{\lambda}\sqrt{n_1^2 \sin^2(\theta_1) - (n_2)^2}$. The electromagnetic waves in the system are written down for $(\theta_1 > \theta_c)$

$$\vec{H}(\vec{r}, t) = \begin{cases} (0, 1, 0)H_I^0 e^{j(kx+\alpha z-\omega t)} + (0, 1, 0)r_p H_I^0 e^{j(kx-\alpha z-\omega t)} & (z \leqslant 0) \\ (0, 1, 0)\frac{n_2}{n_1} t_p H_I^0 e^{-z|\beta|} e^{j(kx-\omega t)} & (z \geqslant 0) \end{cases}$$

$$\vec{E}(\vec{r}, t) = \begin{cases} \left(\frac{\alpha}{\omega}, 0, \frac{-k}{\omega}\right)\frac{H_I^0}{\varepsilon_0\varepsilon_{1r}} e^{j(kx+\alpha z-\omega t)} + \left(\frac{-\alpha}{\omega}, 0, \frac{-k}{\omega}\right) r_p \frac{H_I^0}{\varepsilon_0\varepsilon_{1r}} e^{j(kx-\alpha z-\omega t)} & (z \leqslant 0) \\ \left(j\frac{|\beta|}{\omega}, 0, \frac{-k}{\omega}\right)\frac{H_I^0}{\varepsilon_0\varepsilon_{2r}}\frac{n_2}{n_1} t_p e^{-|\beta| z} e^{j(kx-\omega t)} & (z \geqslant 0) \end{cases}$$

The average magnitude of the Poynting vector of the refractive light in medium 2 only exists in the x-axis direction

$$\begin{aligned} \langle \vec{S} \rangle &= \frac{1}{2} \operatorname{Re}\left\{\vec{E}_T \times \vec{H}_T^*\right\} \\ &= \frac{1}{2} \operatorname{Re}\left\{\left[\left(e^{j\frac{\pi}{2}}\frac{|\beta|}{\omega}\vec{e}_x - \frac{k}{\omega}\vec{e}_z\right)\frac{H_I^0}{\varepsilon_0\varepsilon_{2r}}\frac{n_2}{n_1} t_p e^{-|\beta| z} e^{j(kx-\omega t)}\right]\right. \\ &\quad \left.\times \left[\frac{n_2}{n_1} t_p^* H_I^0 e^{-z|\beta|} e^{-j(kx-\omega t)} \vec{e}_y\right]\right\} \\ &= \frac{|t_p|^2}{2n_1}\sqrt{\frac{\mu_0}{\varepsilon_0}}(H_I^0)^2 e^{-2|\beta| z} \sin(\theta_1)\vec{e}_x \end{aligned}$$

By taking $\cos(\theta_2) = \frac{j}{n_2}\sqrt{n_1^2 \sin^2(\theta_1) - (n_2)^2}$ into formula (3.8), we find the reflectivity of the system is 100%

$$R_p = \left|\frac{E_R^0}{E_I^0}\right|^2 = \left|\frac{n_2 \cos(\theta_1) - j\frac{n_1}{n_2}\sqrt{n_1^2 \sin^2(\theta_1) - (n_2)^2}}{n_2 \cos(\theta_1) + j\frac{n_1}{n_2}\sqrt{n_1^2 \sin^2(\theta_1) - (n_2)^2}}\right|^2 = 1$$

Although there is no average energy flow into medium 2, there is still energy exchange dynamically between the two media across the interface. A refracted light *does* exist in medium 2 even in total reflection, but its amplitude decays with distance perpendicular to the interface of the reflection. The decay distance of the refracted light into medium 2 is defined as the distance where the amplitude of the refracted light is decayed to 37% (= $1/e$). For the ZF7/air system, the decay distance is about 128 nm with $\theta_1 = 45°$ and $\lambda = 632.8$ nm

$$\ell = \frac{\lambda}{2\pi\sqrt{n_1^2 \sin^2(\theta_1) - n_2^2}} \tag{3.11}$$

With total reflection, we can get an evanescent wave traveling along the reflection interface with a higher photon momentum $n_1\hbar\frac{2\pi}{\lambda}\sin(\theta_1)$ which can be used to excite surface plasmons on a metal surface by adjusting the incident angle.

3.1.2 Transverse electric (TE) light

For a monochromatic TE wave shining on an interface between two simple media shown in figure 3.2, the electric field of the incident light is known

$$\vec{E}(\vec{r}, t) = (0, 1, 0)E_I^0 e^{i(kx+\alpha z-\omega t)} \tag{3.12}$$

where $k = n_1\frac{2\pi}{\lambda}\sin(\theta_1)$, $\alpha = n_1\frac{2\pi}{\lambda}\cos(\theta_1)$ and E_I^0 are decided by the incident light direction and light intensity according to formula (2.28). The directions of the reflected light and the refracted light are also set by $\theta_r = \theta_1$ and $\theta_2 = \arcsin[\frac{n_1}{n_2}\sin(\theta_1)]$.

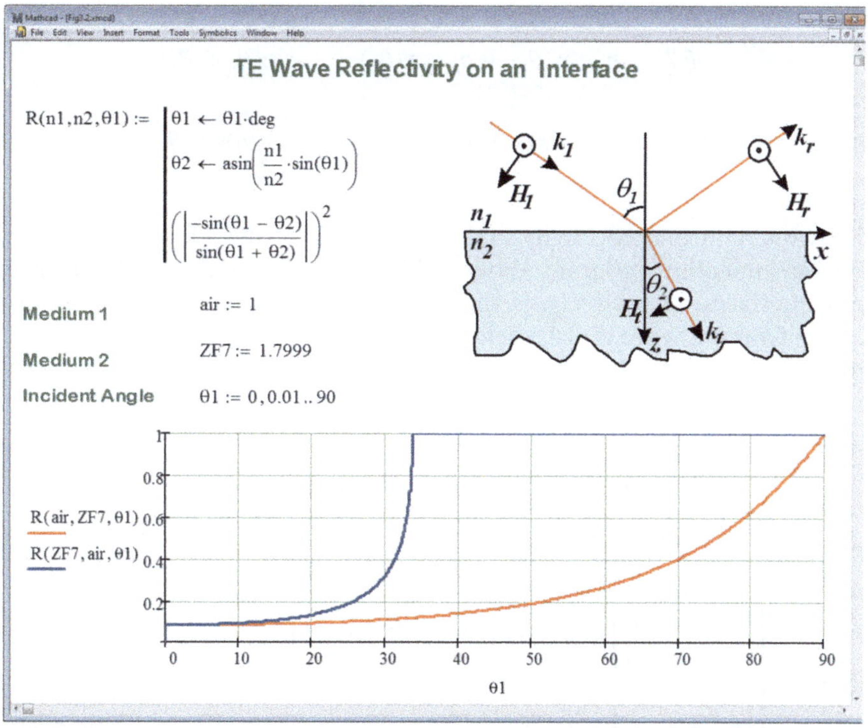

Figure 3.2. A Mathcad program for numerical simulations about a TE wave reflectivity on an air/glass interface (the red curve) and a glass/air interface (the blue curve) with incident angles from 0° to 90°. Available online at http://iopscience.iop.org/book/978-0-7503-2384-0.

The amplitudes of the reflected light E_R^0 and the refracted light E_T^0 can be obtained from the boundary conditions

$$\vec{E}(\vec{r}, t) = \begin{cases} (0, 1, 0)E_I^0 e^{i(kx+\alpha z-\omega t)} + (0, 1, 0)E_R^0 e^{i(kx-\alpha z-\omega t)} & (z \leqslant 0) \\ (0, 1, 0)E_T^0 e^{i(kx+\beta z-\omega t)} & (z > 0) \end{cases} \tag{3.13}$$

With a boundary condition $\vec{n} \times (\vec{E}_2 - \vec{E}_1) = 0$ on the interface ($z = 0$), the tangent components of the electric fields on both sides should be equal

$$E_I^0 + E_R^0 = E_T^0 \tag{3.14}$$

In the system without free current, the magnetic fields in the x direction on the interface are continuous. We have $\vec{n} \times (\vec{H}_2 - \vec{H}_1) = 0$ on the interface. Exactly, it can be written as $-H_I^0 \cos(\theta_1) + H_R^0 \cos(\theta_1) = -H_T^0 \cos(\theta_2)$. With $H = \sqrt{\frac{\varepsilon}{\mu}} E$, the second boundary condition about magnetic field amplitudes is converted to be about electric field amplitudes

$$n_1 E_I^0 \cos(\theta_1) - n_1 E_R^0 \cos(\theta_1) = n_2 E_T^0 \cos(\theta_2) \tag{3.15}$$

Combing the above two equations into a group, we explicitly get its solutions about E_R^0 and E_T^0

$$r_s = \frac{E_R^0}{E_I^0} = \frac{n_1 \cos(\theta_1) - n_2 \cos(\theta_2)}{n_1 \cos(\theta_1) + n_2 \cos(\theta_2)} = -\frac{\sin(\theta_1 - \theta_2)}{\sin(\theta_1 + \theta_2)} \tag{3.16}$$

$$t_s = \frac{E_T^0}{E_I^0} = \frac{2n_1 \cos(\theta_1)}{n_1 \cos(\theta_1) + n_2 \cos(\theta_2)} = \frac{2 \cos(\theta_1)\sin(\theta_2)}{\sin(\theta_1 + \theta_2)} \tag{3.17}$$

Finally, we obtain the reflectivity and transmission of the TE wave in the system. A numeric simulation program about a TE wave reflectivity on the air/glass and glass/air interfaces, respectively, is shown in figure 3.2. The critical angle for total reflection of any TE wave is also decided by equation (3.10). The blue curve in figure 3.2 shows that the critical angle is equal to 33.751°. Although there is an evanescent wave for the TE wave's total reflection, it cannot be used to excite surface plasmons on a metal surface, because eigen-waves are in the TM model

$$\vec{E}(\vec{r}, t) = \begin{cases} (0, 1, 0)E_I^0 e^{i(kx+\alpha z-\omega t)} + (0, 1, 0)r_s E_I^0 e^{i(kx-\alpha z-\omega t)} & (z \leqslant 0) \\ (0, 1, 0)t_s E_I^0 e^{i(kx+\beta z-\omega t)} & (z > 0) \end{cases} \tag{3.18}$$

$$\vec{B}(\vec{r}, t) = \begin{cases} \left(\frac{-\alpha}{\omega}, 0, \frac{k}{\omega}\right) E_I^0 e^{j(kx+\alpha z-\omega t)} + \left(\frac{\alpha}{\omega}, 0, \frac{k}{\omega}\right) r_s E_I^0 e^{j(kx-\alpha z-\omega t)} & (z \leqslant 0) \\ \left(\frac{-\beta}{\omega}, 0, \frac{k}{\omega}\right) t_s E_I^0 e^{j(kx+\beta z-\omega t)} & (z > 0) \end{cases} \tag{3.19}$$

$$R_s = \frac{n_1 \cos(\theta_1)}{n_1 \cos(\theta_1)} |r_s|^2 = \left| \frac{n_1 \cos(\theta_1) - n_2 \cos(\theta_2)}{n_1 \cos(\theta_1) + n_2 \cos(\theta_2)} \right|^2 \tag{3.20}$$

$$T_s = \frac{n_2 \cos(\theta_2)}{n_1 \cos(\theta_1)} |t_s|^2 = \left| \frac{\sqrt{2n_1 n_2 \cos(\theta_1)\cos(\theta_2)}}{n_1 \cos(\theta_1) + n_2 \cos(\theta_2)} \right|^2 \tag{3.21}$$

3.2 Monochromatic light reflection on a thin film between two media

Optical elements, such as lenses and mirrors are usually coated with thin films to modify their responses to incident light for specific wavelengths. For mirrors, we use coatings to increase their reflectivity, and for lenses, we use coatings to increase their transmission. SPR biosensors use thin gold films in their sensing chips to detect biofilms. The reflection and refraction of light waves in a system with a monolayer are also dependent on light polarization mode.

3.2.1 TM light

By adjusting an air gap between a prism and a gold surface, we can use visible light to excite surface plasmon waves. In the optical system shown in figure 3.3, a monochromatic TM wave is shining on a thin film between two media with an incident angle θ in medium 0. The refractive indices of the three media are $n0$, $n1$ and $n2$ respectively, and the thin film's thickness is d. A 3D Cartesian coordinate system is set in the system with its xoy plane on the interface between medium 0 and 1, and the z-axis points from medium 0 to 2. The y-axis is not shown in the figure because it is perpendicular to the xoz plane and points out of the page to readers. At the origin of the coordinate system, a part of the incident light energy is reflected back into medium 0, and a part of the light energy is refracted into the film. There will be a stationary wave inside the film created by multiple reflections and multiple refractions. Some light energy is also refracted into medium 2. The magnetic fields of light waves in the system only have components in the y-axis direction: $\vec{H}(\vec{r}, t) = H \cdot \vec{e}_y$

$$H = \begin{cases} H_{0y} \\ H_{1y} \\ H_{2y} \end{cases} = \begin{cases} [a_0 \cos(\alpha z) + b_0 \sin(\alpha z)] e^{j(kx - \omega t)} & (z \leqslant 0) \\ [a_1 \cos(\beta z) + b_1 \sin(\beta z)] e^{j(kx - \omega t)} & (0 < z \leqslant d) \\ a_2 e^{j(kx + \gamma z - \omega t)} & (z > d) \end{cases} \tag{3.22}$$

In the above expression, a_0, α, β, γ and k are known constants determined by the incident light wave intensity and direction. According to Snell's law with an incident angle θ in medium 0, we can obtain: $k = n_0 \frac{2\pi}{\lambda} \sin(\theta)$, $\alpha = n_0 \frac{2\pi}{\lambda} \cos(\theta)$, $\beta = \frac{2\pi}{\lambda} \sqrt{n_1^2 - n_0^2 \sin^2(\theta)}$, and $\gamma = \frac{2\pi}{\lambda} \sqrt{n_2^2 - n_0^2 \sin^2(\theta)}$.

The other four parameters b_0, a_1, b_1 and a_2 can be obtained from boundary conditions (2.57) of electromagnetic waves on the two interfaces without free electric

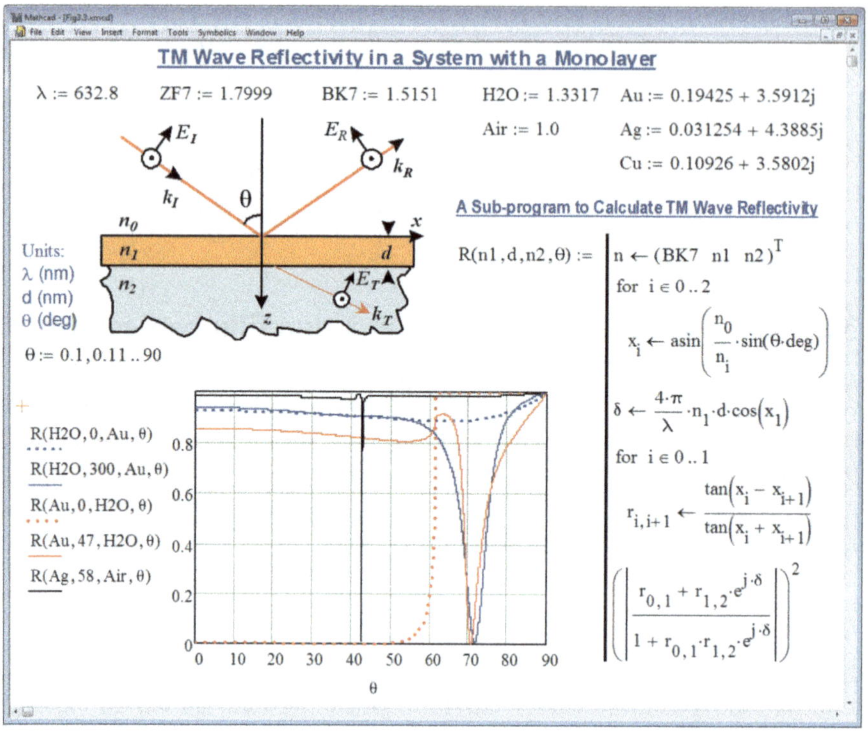

Figure 3.3. A Mathcad program for numeric simulations of reflectivity of a monochromatic TM wave reflected on a thin film between two media. There is a deep valley on each of the three solid curves. These optical adsorption phenomena are caused by surface plasmon resonance. Available online at http://iopscience.iop.org/book/978-0-7503-2384-0.

charges and free currents. It is convenient to define the magnetic field intensities (3.22) in medium 0–2 as H_{0y}, H_{1y} and H_{2y} respectively in the following calculations.

On the two interfaces of the film, the tangent components of magnetic fields are continuous: $(H_{1y} - H_{0y})|_{z=0} = 0$, and $(H_{2y} - H_{1y})|_{z=d} = 0$. Exactly, they could be written as the following two equations

$$a_1 = a_0 \tag{3.23}$$

$$a_2 e^{j\gamma d} = a_1 \cos(\beta d) + b_1 \sin(\beta d) \tag{3.24}$$

In the space without free currents, electric field vectors $\vec{E}$ in the three media could be derived from $\nabla \times \vec{H} = \frac{\partial}{\partial t}\vec{D}$ and $\vec{D} = \varepsilon_0 \varepsilon_r \vec{E}$

$$\vec{D} = \begin{cases} \dfrac{\alpha j}{\omega}[a_0 \sin(\alpha z) - b_0 cos(\alpha z)]e^{j(kx - j\omega t)}\vec{e}_x - \dfrac{k}{\omega}H_{0y}\vec{e}_3 & (z \leqslant 0) \\ \dfrac{\beta j}{\omega}[a_1 \sin(\beta z) - b_1 cos(\beta z)]e^{j(kx - j\omega t)}\vec{e}_x - \dfrac{k}{\omega}H_{1y}\vec{e}_3 & (0 < z \leqslant d) \\ \dfrac{\gamma}{\omega}H_{2y}\vec{e}_x - \dfrac{k}{\omega}H_{2y}\vec{e}_3 & (z > d) \end{cases} \tag{3.25}$$

It is also convenient to define the electric field strength vectors (3.25) in medium 0–2 as $\vec{E}_0$, $\vec{E}_1$ and $\vec{E}_2$ respectively. The tangent components of electric fields are continuous on each side of the two interfaces: $(E_{1x} - E_{0x})|_{z=0} = 0$ and $(E_{2x} - E_{1x})|_{z=d} = 0$. Two more equations are also derived from them by taking equation (3.25) into these boundary conditions

$$\frac{\beta}{\varepsilon_{1r}}b_1 = \frac{\alpha}{\varepsilon_{0r}}b_0 \tag{3.26}$$

$$\frac{\gamma}{\varepsilon_{2r}}a_2e^{j\gamma d} = \frac{j\beta}{\varepsilon_{1r}}[a_1\sin(\beta d) - b_1\cos(\beta d)] \tag{3.27}$$

Equations (3.24) and (3.27) could be combined into a group to find solutions for b_1 and a_2. Then with equations (3.23) and (3.26), each of the four parameters has its explicit expression

$$a_1 = a_0$$

$$b_0 = \frac{\beta}{\alpha}\frac{\varepsilon_{0r}}{\varepsilon_{1r}}b_1$$

$$b_1 = \frac{\frac{-\gamma}{\varepsilon_{2r}}\cos(\beta d) + \frac{j\beta}{\varepsilon_{1r}}\sin(\beta d)}{\frac{\gamma}{\varepsilon_{2r}}\sin(\beta d) + \frac{j\beta}{\varepsilon_{1r}}\cos(\beta d)}a_0$$

$$a_2 = \frac{\frac{j\beta}{\varepsilon_{1r}}e^{-j\gamma d}}{\frac{\gamma}{\varepsilon_{2r}}\sin(\beta d) + \frac{j\beta}{\varepsilon_{1r}}\cos(\beta d)}a_0$$

Taking these parameters into equation (3.22), then replacing $\cos(\theta)$ and $\sin(\theta)$ with $\frac{1}{2}(e^{j\theta} + e^{-j\theta})$ and $\frac{1}{2j}(e^{j\theta} - e^{-j\theta})$ respectively, we could obtain an explicit expression of the magnetic field intensity in medium 0 with a definition $H_I^0 = \frac{1}{2}(a_0 - ib_0)$

$$\begin{aligned} H_{0y} &= \left[a_0\frac{e^{j\alpha z} + e^{-j\alpha z}}{2} + b_0\frac{e^{j\alpha z} - e^{-j\alpha z}}{2j}\right]e^{j(kx-\omega t)} \\ &= \frac{1}{2}[(a_0 - jb_0)e^{j\alpha z} + (a_0 + jb_0)e^{-j\alpha z}]e^{j(kx-\omega t)} \\ &= H_I^0e^{i(kx+\alpha z-\omega t)} + \frac{a_0 + jb_0}{a_0 - jb_0}H_I^0e^{i(kx-\alpha z-\omega t)} \end{aligned}$$

It is clear that the first part on the right of the symbol '=' is the magnetic field of the incident light, and the second part is the composed magnetic field H_R of all reflected light waves in medium 0. The ratio of amplitudes of the reflected light and the incident light can be simplified to a well-known formula.

$$\frac{H_R^0}{H_I^0} = \frac{a_0 + jb_0}{a_0 - jb_0} = \frac{r_{01} + r_{12}e^{j\delta}}{1 + r_{01}r_{12}e^{j\delta}}$$

With some manipulation of mathematics, we obtained an explicit expression about the magnetic field intensity in the system with three media

$$\frac{H}{H_I^0} = \begin{cases} e^{j(kx+\alpha z-\omega t)} + \dfrac{r_{01} + r_{12}e^{j\delta}}{1 + r_{01}r_{12}e^{j\delta}} e^{j(kx-\alpha z-\omega t)} & (z \leqslant 0) \\ \dfrac{n_1}{n_0}\dfrac{t_{01}}{1 + r_{01}r_{12}e^{j\delta}} e^{j(kx+\beta z-\omega t)} + \dfrac{n_1}{n_0}\dfrac{t_{01}r_{12}e^{j\delta}}{1 + r_{01}r_{12}e^{j\delta}} e^{j(kx-\beta z-\omega t)} & (0 < z \leqslant d) \\ \dfrac{n_2}{n_0}\dfrac{t_{01}t_{12}e^{j\delta/2}}{1 + r_{01}r_{12}e^{j\delta}} e^{j[kx+\gamma(z-d)-\omega t]} & (z > d) \end{cases} \quad (3.28)$$

Inside the expression, r_{01}, r_{12}, t_{01}, and t_{12} are Fresnel coefficients of TM waves on the two interfaces of the film, as shown in equations (3.6)–(3.7). It is well known that the optical phase difference of two neighboring reflected light waves in medium 0 is:

$$\delta = \frac{4\pi}{\lambda} d\sqrt{n_1^2 - n_0^2 \sin^2(\theta_0)} \quad (3.29)$$

To find the reflectivity of the system with a film to a TM wave, we need to know the ratio of the amplitudes of electric fields between the composed reflective wave and the incident wave

$$r_p = \frac{E_R^0}{E_I^0} = \frac{\sqrt{\dfrac{\mu_o\mu_{0r}}{\varepsilon_o\varepsilon_{0r}}} H_R^0}{\sqrt{\dfrac{\mu_o\mu_{0r}}{\varepsilon_o\varepsilon_{0r}}} H_I^0} = \frac{H_R^0}{H_I^0} = \frac{r_{01} + r_{12}e^{j\delta}}{1 + r_{01}r_{12}e^{j\delta}} \quad (3.30)$$

$$R_p = \left| \frac{E_R^0}{E_I^0} \right|^2 = \left| \frac{r_{01} + r_{12}e^{j\delta}}{1 + r_{01}r_{12}e^{j\delta}} \right|^2 \quad (3.31)$$

Also, we have a ratio of the amplitudes of electric fields between the composed refractive light at the interface of $z = d$ and the incident light at the interface of $z = 0$

$$t_p = \frac{E_T|_{z=d}}{E_I|_{z=0}} = \frac{\sqrt{\dfrac{\mu_o\mu_{2r}}{\varepsilon_o\varepsilon_{2r}}} H_T^0}{\sqrt{\dfrac{\mu_o\mu_{0r}}{\varepsilon_o\varepsilon_{0r}}} H_I^0} e^{j\beta d} = \frac{n_0}{n_2}\frac{H_T^0}{H_I^0} e^{j\beta d} = \frac{t_{01}t_{12}e^{j\delta/2}}{1 + r_{01}r_{12}e^{j\delta}} \quad (3.32)$$

The electric field strength of the composed refractive light in medium 2 is explicitly written below. Its direction is decided by k and γ in the xoz plane shown in figure 3.3

$$E_T(\vec{r}, t) = \frac{t_{01}t_{12}}{1 + r_{01}r_{12}e^{j\delta}} e^{j\beta d} E_I^0 e^{j[kx+\gamma(z-d)-\omega t]} \tag{3.33}$$

In total reflection, the transmitted light is an evanescent wave in medium 2 because both β and γ are complex numbers.

A Mathcad program for numeric simulations of a TM light wave ($\lambda = 632.8$ nm) reflected in a system with a monolayer is shown in figure 3.3. The first line of the program is a text region giving the title of the program. The following three lines list refractive indices of two glasses (ZF7, BK7), two dielectrics (water and air), and three metals (gold, silver and copper) for the monochromatic light. These parameters come from a free database published by Mikhail Polyanskiy on the website https://refractiveindex.info. Each refractive index number is assigned to a variable with the abbreviated name of the material. A schematic about the structure of the system and light polarization was created in Microsoft PowerPoint, and then pasted in the top-left corner of the program.

Line 5 of the program is also a text region for the title of a sub-program or function $R(n1, d, n2, \theta)$ to return reflectivity of the TM light in medium 0. There are four local variables in the function. They are the thin film's refractive index, the film thickness, the refractive index of medium 2, and the incident angle of the TM light in medium 0. Glass BK7 is set to be medium 0 by the first line in the function. In Otto configuration for SPR excitation, medium 1 is a dielectric film, and medium 2 is a smooth solid metal. But in Kretschman configuration for SPR excitation, medium 1 is a thin metal film, and medium 2 is a dielectric. In both configurations, medium 0 is a glass for total internal reflections.

According to Snell's law, a loop in line 2 and 3 of the function calculates angles in the three media, and the unit of the output angle x_0, x_1 and x_2 is radians, but the unit of the incident angle θ is degrees. Line 4 in the function calculates the phase difference δ according the formula (3.29). With a loop set by lines 5 and 6, Fresnel coefficients of reflected light on the two interfaces are calculated according to the formula (3.6), and finally the reflectivity of the system can be obtained to $R(n1, d, n2, \theta)$ when it is called. An x–y plot in the program is the output of the program. The two blue curves are for Otto configuration, the black and two red curves are for Kretschman configuration. The dashed blue curve is from $R(H_2O, 0, Au, \theta)$. Since the water film thickness is set to be zero, there are only two media in the system with only one interface between BK7 glass and gold. The system is just a golden mirror with reflectivity over 90%. If the glass and the gold were separated by a water film with a thickness of 300 nm by $R(H_2O, 300, Au, \theta)$, the system is still a good mirror, but in the case of $60° \leqslant \theta \leqslant 80°$, there is an adsorption valley. This is a surface plasmon resonance spectrum with a resonance angle θ_{SPR} at 72°. It is a real device for invisibility but only for incident angles around 72°.

The dashed red curve was drawn by calling of $R(Au, 0, H_2O, \theta)$. There is no gold film in the system as the gold film thickness was set to be zero in the function. The Brewster angle and the critical angle of the system with BK7 glass and water is 41.314° and 61.516° respectively. The solid red curve is a SPR spectrum with a gold film of 47 nm between BK7 and water. The resonance angle θ_{SPR} is smaller than that

Table 3.1. Otto configuration θ_{SPR}.

Prism	Gap	Gold		Silver		Copper	
n_0	ε_{1r}	$Re(\varepsilon_{2r})$	$\theta_{SPR}(°)$	$Re(\varepsilon_{2r})$	$\theta_{SPR}(°)$	$Re(\varepsilon_{2r})$	$\theta_{SPR}(°)$
1.5150	1.3317^2	−12.859	71.19	−19.258	67.28	−12.806	71.24
1.5150	1		43.42		42.68		43.43
1.7999	1.3317^2		52.82		50.93		52.85
1.7999	1		35.35		34.79		35.35

of the Otto configuration. If we use ZF7 to replace BK7 in the function $R(n1, d, n2, \theta)$ definition, the SPR spectra will be so sharp that it will not be easy to see the resonance angles' difference of the two configurations in the x–y plot. The solid black curve is the SPR spectrum of the system with three media: BK7 glass, a silver film of 58 nm and air. It is so sharp that we can only see a vertical straight line at 42°. Copper can also produce sharper SPR spectra. Gold is suitable for biosensors because of its noble properties.

According to formula (2.76), we could use the following formula to guess the surface plasmon resonance angle θ_{SPR}. Table 3.1 shows some θ_{SPR} values calculated from equation (3.34) for Otto configuration. The wavelength was set to be 632.8 nm in the calculations. In Kretschman configuration, there are two different dielectrics attached to the metal film. This system is very complex without an explicit expression for its resonant angles. But we could still use this formula to guess θ_{SPR} for a system with Kretschman configuration

$$\theta_{SPR} = \arcsin\left(\frac{1}{n_0}\sqrt{\frac{\varepsilon_{1r}\varepsilon_{2r}'}{\varepsilon_{1r} + \varepsilon_{2r}'}}\right) \tag{3.34}$$

3.2.2 TE light

For the monochromatic TE wave shining on a film between two media shown in figure 3.4, the electric fields of the system only have components in the y-axis direction: $\vec{E}(\vec{r}, t) = E \cdot \vec{e}_y$. Explicit expressions of electric fields were obtained by using $\nabla \times \vec{E} = \frac{\partial}{\partial t}\vec{B}$ and two boundary conditions on the two interfaces of the system without free currents for the tangent components of $\vec{E}$ and $\vec{H}$. The constant E_I^0 in the expression was set by the incident light intensity

$$\frac{E}{E_I^0} = \begin{cases} e^{j(kx+\alpha z-\omega t)} + \dfrac{r_{01} + r_{12}e^{j\delta}}{1 + r_{01}r_{12}e^{j\delta}} e^{j(kx-\alpha z-\omega t)} & (z \leqslant 0) \\[2ex] \dfrac{t_{01}}{1 + r_{01}r_{12}e^{i\delta}} e^{j(kx+\beta z-\omega t)} + \dfrac{t_{01}r_{12}e^{i\delta}}{1 + r_{01}r_{12}e^{i\delta}} e^{j(kx-\beta z-\omega t)} & (0 < z \leqslant d) \\[2ex] \dfrac{t_{01}t_{12}e^{j\delta/2}}{1 + r_{01}r_{12}e^{j\delta}} e^{j[kx+\gamma(z-d)-\omega t]} & (z > d) \end{cases} \tag{3.35}$$

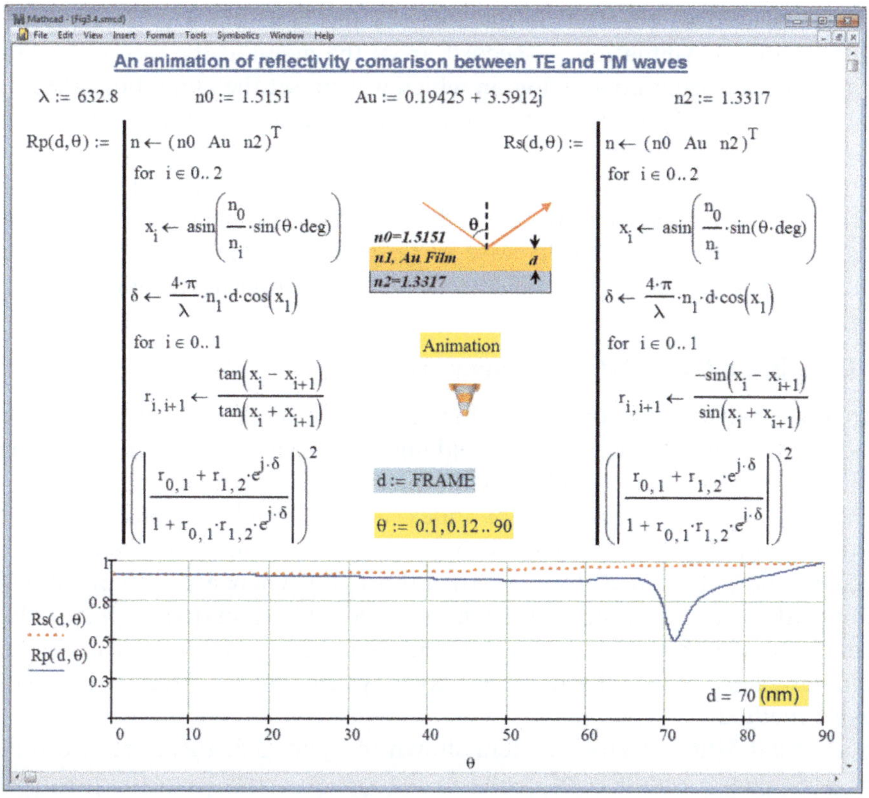

Figure 3.4. A Mathcad program used to make and show a video for reflectivity comparison between TE and TM waves reflected on a thin gold film between BK7 glass and water. The thickness of the film set by FRAME changes from 0 to 100 nm. Video and program available online at http://iopscience.iop.org/book/978-0-7503-2384-0.

Although the reflectivity formula of the TE wave in the system is the same as formula (3.31), Fresnel coefficients r_{01}, r_{12}, t_{01} and t_{02} inside equation (3.35) should be the formulae shown in equations (3.16)–(3.17) respectively for the TE wave

$$R_s = \left| \frac{E_R^0}{E_I^0} \right|^2 = \left| \frac{r_{01} + r_{12}e^{j\delta}}{1 + r_{01}r_{12}e^{j\delta}} \right|^2 \tag{3.36}$$

In the Mathcad program shown in figure 3.3, you can replace $\frac{\tan(x_i - x_{i+1})}{\tan(x_i + x_{i+1})}$ with $\frac{-\sin(x_i - x_{i+1})}{\sin(x_i + x_{i+1})}$ to see the reflectivity of a TE wave in that system. No SPR spectra can be seen in the *x*–*y* plot of that program for the TE wave, although total internal reflections still happen. A video associated with figure 3.4 shows a significant difference of reflectivity between TE and TM waves with the thickness of a gold film varying from 0 to 100 nm. In the program, *Rp*(*d*, *θ*) and *Rs*(*d*, *θ*) are sub-programs defined for the reflectivity of TM and TE waves respectively. The local variable *d* is the thickness of the thin gold film. Mathcad has a predefined constant called FRAME which is used to drive animations. In the program, the animation parameter is used to control the thickness of the gold film by 'd:=FRAME'. By clicking 'Tools → Animation → Record' from the menu at the top of the Mathcad interface, we can bring up a dialog, in which we can set the animation frame from 0 to 100 in a speed of 10 frames per second. Then we drag-select the region of the x–y plot and press 'Animate' on the interface of the dialog to create an AVI format video file with a size of 1.7 Mb. The AVI file is then converted into a MP4 format video file (0.153 Mb) by a free media player VLC. The smaller video file is then inserted into the Mathcad file as an object. A light wave in TE model is useless for SPR, but it is very useful as a reference for light phase detections in other optical technologies.

3.3 Benchmarks of SPR biosensors

Up to now, we handled only one thin film (gold) deposited on a prism in our Mathcad programs. Practically each gold film is deposited on a microscope coverslip (sizes: $20 \times 20 \times 0.2\ mm^3$) made of BK7 glass. After a coverslip with a gold film is attached to a prism by a drop of index matching oil, receptor molecules are immobilized on the gold film to form a second film in water; and then ligand molecules bind to their receptors to form a third film. There are so many thin films in each optical biosensing system, it is almost impossible to find an explicit reflectivity formula by solving Maxwell's equations with boundary conditions on each interface. Fortunately, light reflectivity in a multilayer system can be calculated by Rouard's recursive method [2].

In a virtual SPR biosensor system shown in figure 3.5, there are two thin films ($N = 2$). Sequence numbers of media in the system starts from the prism with an integer number $i = 0$. Since the prism and the coverslip are treated as a whole if the prism is also made of BK7 glass, the gold film's sequence number is $i = 1$, and so on. For a TM monochromatic light wave shining on the interface between medium 0 and 1 with an incident angle θ, Fresnel coefficients on each interface are known: $r_{i,i+1}$ ($i = 0, 1\ldots N$). Using Rouard recursive method, we can start from the last thin film by treating its two interfaces as an equivalent interface which would contact with its neighbor film ($N - 1$)

$$r_{N-1,N+1} = \frac{r_{N-1,N} + r_{N,N+1}e^{j\delta_N}}{1 + r_{N-1,N}r_{N,N+1}e^{j\delta_N}} \tag{3.37}$$

In this way from the bottom to the prism (medium 0), all films are integrated into an equivalent interface between medium 0 and medium ($N + 1$)

$$r_{i,N+1} = \frac{r_{i,i+1} + r_{i+1,N+1}e^{j\delta_{i+1}}}{1 + r_{i,i+1}r_{i+1,N+1}e^{j\delta_{i+1}}} \quad (i = N - 1,\ N - 2,\ \ldots 0) \tag{3.38}$$

The Mathcad program shown in figure 3.5 is a practice of the Rouard recursive method for a virtual SPR biosensor. In the second line of the program, optical parameters are assigned to five variables for the virtual optical system shown on the bottom-left corner of the figure. The sample film parameters ($n2$, $d2$) and the incident

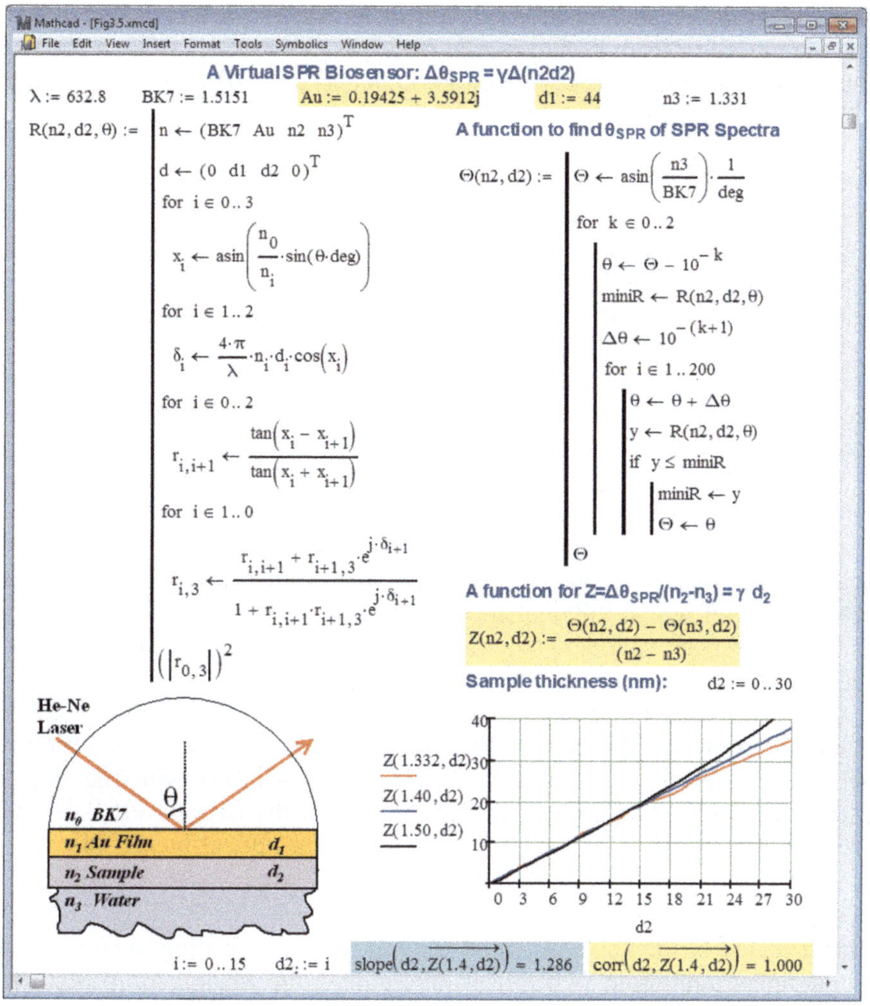

Figure 3.5. A Mathcad program for numerical simulations of a virtual SPR biosensor. The sub-program $R(n2, d2, \theta)$ is used to generate SPR spectra. A function $\Theta(n2, d2)$ can automatically find θ_{SPR} of any SPR spectrum. The x–y plot presents a linear relationship: $\Delta\theta_{\mathrm{SPR}} = \gamma\cdot\Delta(d2n2)$, where $d2 \leqslant 15$ nm and $\gamma = 1.286$. Available online at http://iopscience.iop.org/book/978-0-7503-2384-0.

angle (θ) in degrees are local variables for a sub-program or function $R(n2, d2, \theta)$ defined in the third line of the program. It is very similar to the Mathcad program shown in figure 3.3, but the last loop based on equation (3.38) in the function is new. Since there are only two thin films in the system, the loop starts from $N - 1 = 1$. On the right of the fourth line of the program, the function $\Theta(n2, d2)$ is defined. It is used to search and find the resonance angle θ_{SPR} of any SPR spectrum. The search starts from the critical angle of the system: $\theta_c = \arcsin(n3/\mathrm{BK7})$. In the search procedure, the function $R(n2, d2, \theta)$ would be called 3×200 times.

As the incident angle unit in $R(n2, d2, \theta)$ is degrees, a factor 1/deg is used to convert the unit of θ_c from radians to degrees. It is 61.5°. Then a major loop with an integer k begins a search process from an incident angle ($\theta_c - 10^{-k}$) with a step of $\Delta\theta = 10^{-(k+1)}$ degrees. For example in the beginning, $k = 0$, the initial incident angle is 60.5°, and the searching step is 0.1°. The reflectivity of the system at ($\theta_c - 1$) is obtained by calling $R(n2, d2, \theta)$, and its value is assigned to a variable named miniR. An internal loop with an integer i will increase the incident angle value by one step ($\Delta\theta$), it then calls the function $R(n2, d2, \theta)$ by '$y \leftarrow R(n2, d2, \theta)$'. If the value of y is less than or equal to the value of miniR, the value of miniR will be updated and the incident angle corresponding to miniR will be recorded. By the end of the internal loop, θ_{SPR} will be found with a precision of $10^{-(k+1)}$ degrees. In the function $k = 2$, we can locate θ_{SPR} with an error within 0.001° in the numeric simulations.

Above an x–y plot of the output of the program, one more function $Z(n2, d2)$ is defined. It is about the ratio between $\Delta\theta_{SPR}$ and ($n2 - n3$). The x–y plot presents three curves for $n2 = 1.332$, 1.4 and 1.5 respectively. In chapter 1, we made a hypothesis about protein adsorptions on a gold surface. During an adsorption process for protein, the thickness of the sample is equal to the protein molecule's diameter which will not change. Only the surface concentration of the protein changes, so the refractive index of the sample layer over the gold film can increase from $n3$ (water refractive index) to a maximum of 1.5 when the entire gold surface of a sensing chip is fully occupied. We can see that there is a linear relationship between $Z(n2, d2)$ and $d2$ if $d2 \leqslant 15$ nm. The slope of the linear relationship is $\gamma = 1.286$ [deg nm^{-1}] and the correlation coefficient is 1.000. Based on these numerical simulation results, we get a linear relationship about $\Delta\theta_{SPR}$ and $(n2 - n3)d2$ with the slope value $\gamma = 1.286$ [deg nm^{-1}]. The slope is the benchmark for the virtual SPR biosensor. We will use the benchmark γ for comparisons with other optical biosensors

$$\Delta\theta_{SPR} = 1.286\Delta(n2 \cdot d2) \tag{3.39}$$

The linear region of the above formula is up to 3.26°. According to equation (1.12), we get a linear relationship between $\Delta\theta_{SPR}$ and the change of protein surface concentration $\Delta\Gamma$ in a unit of [ng mm^{-2}]

$$\Delta\theta_{SPR} = 0.23\Delta\Gamma \tag{3.40}$$

The linear region of the virtual SPR biosensor is found to be up to 14.2 ng mm^{-2}. Is the formula from numeric simulations reliable? In 1991, Stenberg *et al* published a paper [3] where optical parameters of their SPR biosensor (BIAcore 1000) were listed. The wavelength of their light source was $\lambda = 760$ nm. At this wavelength, a BK7 prism's refractive index was 1.511, the gold film's thickness was 45 nm with a refractive index $0.17 + j4.93$, and water refractive index was 1.3298. Taking these parameters into the Mathcad program shown in figure 3.5, the benchmark of the system is found to be $\gamma = 0.61$ [deg nm^{-1}] in a condition of $d3 \leqslant 26$ nm or $\Delta\theta_{SPR} \leqslant$ 2.7°, or $\Delta\Gamma \leqslant 24.5$ nm mm^{-2}

$$\Delta\theta_{SPR} = 0.11\Delta\Gamma \tag{3.41}$$

In their experimental data reported in the paper, surface concentrations of ^{14}C-labeled chymotrypsinogen, ^{14}C-labeled transferrin, and ^{35}S-antitransferrin monoclonal antibodies were directly measured for comparisons with their SPR system's responses ($\Delta\theta_{SPR}$). From a best-fitting of a linear regression, a linear relationship was presented in the paper. Corresponding to 1.0 ng mm^{-2} protein adsorbed on a gold surface, their SPR biosensor's resonance angle shifted 0.10° ± 0.01°. This linear relationship was valid for protein surface concentrations between 2 to 50 [ng mm^{-2}], and its correlation coefficient was 0.975. The numeric simulation formula (3.41) matches with the upper limit of the experimental linear relationship. However, the numeric simulation's linear region is only half of the experimental linear region. Only if we reduce the correlation coefficient from 1.000 to 0.993 for all data points on the three curves shown in the lower-right corner of figure 3.5, is it possible to extend the linear region of the numeric simulation formula to 53 nm to match with that of their experimental formula.

References

[1] Otto A 1968 A new method for exciting non-radioactive surface plasma oscillations *Phys. Stat. Sol.* B **26** K99–K101

[2] Heavens O S 1991 *Optical Properties of Thin Solid Films* (New York: Dover)

[3] Stenberg E, Persson B, Roos H and Urbaniczky C 1991 Quantitative determination of surface concentration of protein with surface plasmon resonance using radiolabeled proteins *J. Collioid Interface Sci.* **143** 513–26

IOP Publishing

Super Optical Biosensors

Caide Xiao

Chapter 4

SPR ellipsometry biosensors

Usually, a linear polarized light reflected on interfaces of a thin film becomes an elliptical polarized light. Using a null ellipsometer, we can measure a phase angle and an amplitude angle of the reflection, and from these angles it is possible to find the thickness and the refractive index of the film. The principle, structure and numerical simulations of a null ellipsometer are presented.

Combining SPR with ellipsometry, we can build a super optical biosensor SPREP, whose benchmark can be tuned by the light incident angle, thicknesses of chrome and gold films. The Levenberg–Marquardt algorithm is used in Mathcad programs to interpret SPREP spectra.

Gold is used in SPR biosensors for its noble properties. With a chemical reaction between a thiol group (–SH) and a gold atom, it is easy to attach receptor molecules with thiol groups to gold surfaces of sensing chips. From SPR theory and numeric simulations in the last chapter, we see that the thickness and the refractive index of a gold film are critical to SPR spectra. Gold refractive indices could be found from many sources [1], such as the database in the website https://refractiveindex.info and the *CRC Handbook of Chemistry and Physics* [2], but those data are very different from each other. The refractive index of a gold film is not only sensitive to its thickness but also to its microstructures. Usually, thin gold films are deposited on glass by thermal evaporation in vacuum. Fine gold particles deposited on glass form a thin film. Gold atoms in the thin film re-crystallize to form microstructures which are dependent on evaporation conditions. Can we directly measure thin gold films used in SPR sensing chips to find their thicknesses and refractive indices?

The answer is 'yes', if we use an optical instrument invented by Paul Drude [3] in 1887. Alexandre Rothen [4] in 1945 gave a name to the instrument: ellipsometer. In 1957 he improved the sensitivity and reliability of the optical instrument by using a photomultiplier tube [5] to detect weak light for null ellipsometers. The non-destructive and non-contact optical technique is called ellipsometry (EP) which has the capability to measure a film's thickness from 0.3 Å to a few microns, and the refractive index of the

doi:10.1088/2053-2563/ab0823ch4

film can be complex numbers. In the 1950s, Nick Bashara and Rasheed Azzam [6] developed methods to use ellipsometry for quality control of semiconductor materials. Ellipsometers are still useful instruments in the semiconductor industry. In 1947, Rothen [7] used an ellipsometer to measure antigen–antibody interactions. Since then, ellipsometers have been used for biological applications [8–13]. However, ellipsometry can only directly measure phase angle and amplitude angle information about two polarized light waves reflected from samples. To find thicknesses and refractive indices information about a sample, we have to choose the right physical model for our sample and do complex mathematical calculations. Only after computers were invented could ellipsometers become more useful and popular.

4.1 Phase angle and amplitude angle of reflected light waves

In the optical system shown in the lower-left corner of figure 4.1, there is a gold film deposited on a coverslip made of BK7 glass in the air. This is a model of the sample

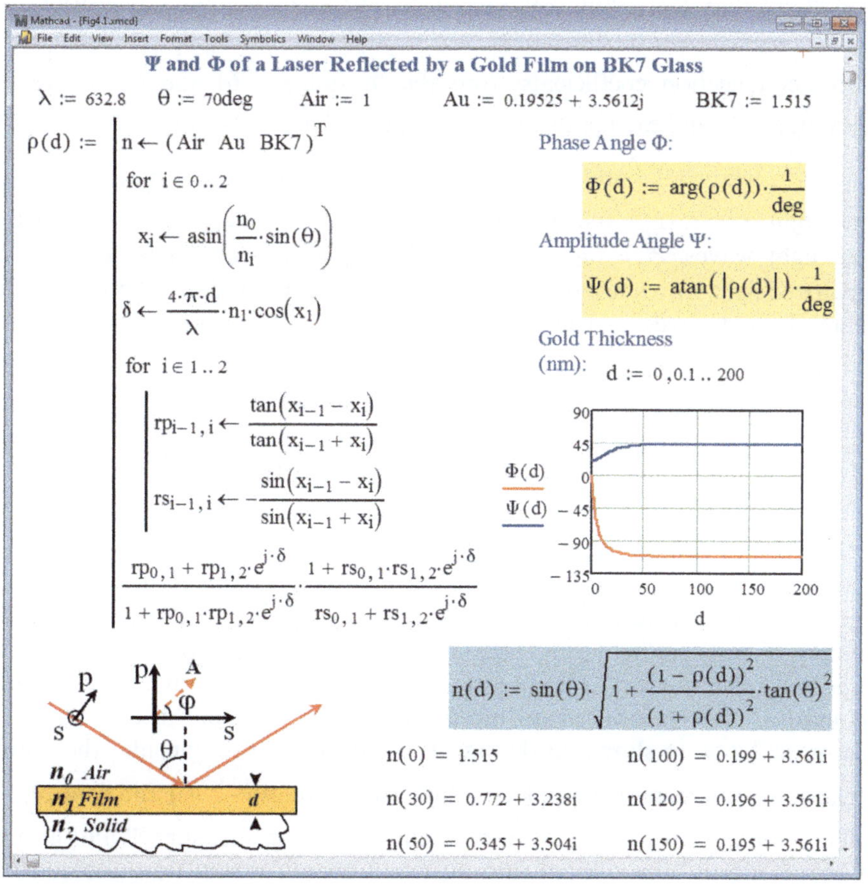

Figure 4.1. Mathcad program for numeric simulations of ellipsometry about the amplitude angle Ψ and the phase angle Φ changed with the thickness of a gold film deposited on BK7 glass in the air. Available online at http://iopscience.iop.org/book/978-0-7503-2384-0.

system with medium 0, 1 and 2, and they are air, a thin gold film and a bulk of BK7 glass respectively. A collimated linear polarized laser ($\lambda = 632.8$ nm) propagating in the air strikes an interface between medium 0 and 1 with an incident angle θ (= 70°). The incident light and the normal of the interface (the black dashed line) construct an incident plane on the page. Another plane constructed by a p-axis and an s-axis is perpendicular to the incident light ray. The two p-axes in the figure are the same. The symbol '⊗' indicates that the s-axis is perpendicular to the incident plane and points into the page. The red dashed line with an arrow represents the electric field amplitude of the incident light: $E = Ae^{j(kr-\omega t)}$, where $k = n_0\frac{2\pi}{\lambda}$ is the length of the light wavenumber vector, r is the traveling distance of the light along the light path at time t, and ω is the angular frequency of the monochromatic light.

The polarization plane of the light is a plane with the incident light ray and the electric field vector. The angle between the incident plane and the polarization plane is 45°, so the angle between the polarization plane and the s-axis is also 45° (φ). The incident light could be decomposed into an s-polarized light and a p-polarized light just before they are reflected by the film: $E_s = A\cos(\frac{\pi}{4})e^{j(kr-\omega t)}$ and $E_p = A\sin(\frac{\pi}{4})e^{j(kr-\omega t)}$. The Fresnel reflection coefficients from the formulae (3.6) and (3.16) are usually complex numbers, and each of them can be expressed in a polar form: $r_p = \frac{|E_{pr}|}{|E_{pi}|}e^{j\phi_p}$ and $r_s = \frac{|E_{sr}|}{|E_{si}|}e^{j\phi_s}$, where $|E_{pr}|$ and $|E_{sr}|$ are the amplitudes of the electric fields of the two reflected light waves, $|E_{pi}|$ and $|E_{si}|$ are the amplitudes of the electric fields of the two incident light waves, $\phi_p = (\phi_r - \phi_i)_p$ and $\phi_s = (\phi_r - \phi_i)_s$ are phase changes of the two polarized light waves respectively. With the same amplitudes and phases of the two incident polarized waves, we can simplify the ratio between r_p and r_s, and define it as an ellipsometry ratio:

$$\rho = \tan(\Psi)e^{j\Phi} = \frac{r_p}{r_s} = \frac{\frac{|E_{pr}|}{|E_{pi}|}e^{j(\phi_r-\phi_i)_p}}{\frac{|E_{sr}|}{|E_{si}|}e^{j(\phi_r-\phi_i)_s}} = \frac{|E_{pr}|}{|E_{sr}|}e^{j(\phi_p-\phi_s)_r} \tag{4.1}$$

In the above formula about the reflection of the two polarized light waves, the amplitude angle is Ψ whose tangent function value is equal to the amplitude ratio between the p-polarized light and the s-polarized light, and the phase angle is Φ, the phase difference between the p-polarized light and the s-polarized light. Traditionally, the symbol 'Δ' is used to represent the phase angle. But this symbol (Δ) is used as the symbol for an increment of any variable in this book. In the simple optical system shown in the figure, we only need to consider the amplitude and phase information of the reflected light: $\Psi = \arctan\left(\frac{|E_{pr}|}{|E_{sr}|}\right)$ and $\Phi = (\phi_p - \phi_s)_r$. A null ellipsometer is an optical instrument which can measure Ψ and Φ directly. With the model of the optical film system under investigation, we can extract information about the optical system: the refractive index and the thickness of the film. The two parameters (Ψ and Φ) are independent of the configuration of the ellipsometer.

The Mathcad program shown in figure 4.1 is used to simulate ellipsometry in which Ψ and Φ were changed with the thickness (d) of a gold film deposited on BK7 glass in the air. The first line in the program is a text region for the title of the program. Five variables are assigned with numbers, respectively in the second line. They are the wavelength (λ = 632.8 nm), the incident angle (θ = 70°), and the refractive indices of the three media: air, gold, BK7 glass. In the third line of the program the function $\rho(d)$ is defined with the thickness of the gold film as its local variable. The first line in the function is used to assign the refractive indices of the three media to a vector named 'n'. By a loop in the second and third line of the function, the reflection angle and two refraction angles are calculated and assigned to a vector named 'x'. The unit of each element in the vector x is in radians. The fourth line of the function calculates the phase angle shift δ contributed by the gold film. Another loop with three lines in the function calculates four Fresnel coefficients on the two interfaces of the film. Two of them are for the p-polarized light and the other two are for the s-polarized light. The last line of the function returns the ellipsometry ratio between r_p and r_s. Using a Mathcad function arg(z), a sub-program is defined for the phase angle Φ whose unit is converted from radians into degrees by a constant 1/deg. For example, if $z = 1 - 1j$, then the value of arg(z)/deg is equal to −45°. The value of Φ is in the region from −180° to 180°. One more sub-program is also defined to calculate the amplitude angle Ψ in degrees. An x–y plot on the right of the figure is the output of the program with the gold film's thickness changing from 0 to 200 nm.

In the case when the thickness of the gold film is equal to zero, there are only media 0 and 2 in the system with only one interface between them. Using symbols $\overline{\Psi}$ and $\overline{\Phi}$ for the amplitude angle and phase angle of the simplified system respectively, we can obtain a complex number $\overline{\rho} = \tan(\overline{\Psi})e^{j\overline{\Phi}}$ from measurements. Then by replacing r_p with $\frac{n_2\cos(\theta) - n_0\cos(\theta_2)}{n_2\cos(\theta) + n_0\cos(\theta_2)}$ and replacing r_s with $\frac{n_0\cos(\theta) - n_2\cos(\theta_2)}{n_0\cos(\theta) + n_2\cos(\theta_2)}$ in $\overline{\rho} = r_p/r_s$, we get n_2/n_0 by an explicit formula [14] shown in equation (4.2). In order to derive the formula, we have to simplify the square of the ratio between $(1 - \overline{\rho})$ and $(1 + \overline{\rho})$ first, and then replace $\cos^2(\theta_2)$ with $1 - \frac{n_0^2}{n_2^2}\sin^2(\theta)$ in the ratio

$$\frac{n_2}{n_0} = \sin(\theta)\sqrt{1 + \left(\frac{1 - \overline{\rho}}{1 + \overline{\rho}}\right)^2 \tan^2(\theta)} = \tan(\theta)\sqrt{1 - \frac{4\overline{\rho}}{(1 + \overline{\rho})^2}\sin^2(\theta)} \tag{4.2}$$

In the lower-right corner of the Mathcad program, the sub-program $n(d)$ is defined according to the above formula with $n_0 = 1$. It is expected that $n(0) = n_2$, and this is the case. If the gold film is thick enough, we expect $\rho = \frac{rp01}{rs01}$ and $\rho(\infty) = n_1$. The x–y plot in the program shows that even a gold film can be treated as bulk gold if its thickness is more than 100 nm. The calculations on the lower-right corner of the program confirm that the formula (4.2) can be used to guess the refractive indices of gold films. No matter what thickness the gold film is, using formula (4.2) could find a complex number for the gold film's refractive index. Although the real part values of the complex numbers are bigger than the true value (0.195 25), the imaginary part values are very close to the true value (3.5612) of the gold film.

4.2 The principle of a null ellipsometer

Our eyes cannot see light phase information, but we can use a quarter-wave plate ($\lambda/4$ plate, or compensator) and two polarizers to convert phase angle signals into light intensity signals in the 'null ellipsometer' which was invented by Drude [15]. A configuration of polarizer–compensator–sample–analyzer (PCSA) is usually used in null ellipsometers. In the x–y plot shown in figure 4.1 about a linear polarized light reflection, the reflected TM wave and TE wave have a phase difference Φ around $-100°$ for a gold film of 200 nm. The superposition of the two reflected waves is an elliptical polarized light. The phase difference (Φ) is an unknown number set by the sample under investigation. By inserting a $\lambda/4$ plate between the polarizer and the sample, we created an elliptical polarized incident light. The phase angle difference between the TM wave and TE wave in the elliptical polarized incident light is adjustable by rotating the polarizer. If we adjusted the incident phase difference to $-\Phi$ or $(180° - \Phi)$, the phase difference of the reflected TM wave and TE wave was compensated to be 0° or 180°, so that the composed reflected light becomes a linear polarized wave which is easy to detect by rotating the analyzer. A photomultiplier receives the maximum light intensity if the principle section of the analyzer is parallel with the polarization plane of reflected light, or the photomultiplier receives a minimum light intensity if the two planes are perpendicular to each other. Since our eyes and instruments are more sensitive in the dark, we rotate the analyzer to get a null reflection of the system for more accurate measurements. That is why this kind of optical instrument is called a 'null ellipsometer'.

4.2.1 The structure of a null ellipsometer

A schematic of a null ellipsometer is shown in figure 4.2. A film deposited on top of a solid is horizontally put on the sample stage of the instrument. A vertical incident plane is constructed to the normal of the film (the black dashed line) and the incident light beam. A semi-circular cast iron frame with two Vernier scales is used to support two arms for optical elements. The rotation axis of the two arms is a virtual line through the reflection point on the sample, and the line is perpendicular to the incident plane. The incident angle (θ) of the light is set by rotating the two arms symmetrically around the reflection axis. In the left arm, there is a collimated He–Ne laser tube, a Glan–Taylor prism polarizer and a $\lambda/4$ plate. An un-polarized laser emitted by the laser tube has polarization planes randomly in all directions.

The polarizer converts the laser into a linear polarized light wave. The polarization plane of the linear polarized light is set by the Glan–Taylor prism's principle section which is rotatable in an s–p Cartesian coordinate system shown on the lower-left corner of the figure. The s–p plane of the Cartesian coordinate is perpendicular to the laser beam with its p-axis in the incident plane and the s-axis perpendicular to

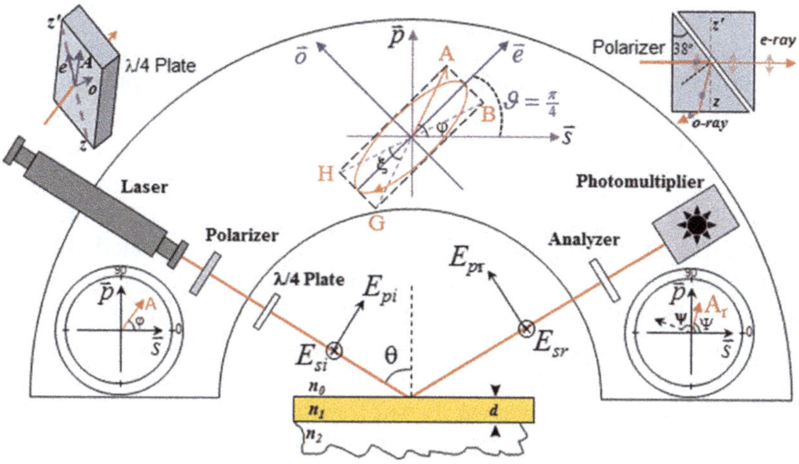

Figure 4.2. A schematic of a null ellipsometer with PCSA configuration. Two rotatable arms hold optical components: a collimated He–Ne laser, a polarizer, a $\lambda/4$ plate, an analyzer and a photomultiplier tube. The structures of the $\lambda/4$ plate and the polarizer are shown on the two top corners respectively. They are made of calcite crystal with their optical axes shown by pink dashed lines. The double coordinate systems on top are located at the reflecting point with its p-axis parallel with another p-axis on the left. The sample model is a monolayer sandwiched by medium 0 and 2.

the incident plane. The symbol '⊗' on the laser beam indicates that the s-axis points into the page perpendicularly. The length of the arrow (A) represents the amplitude of the electric field of the linear polarized light, and the angle (φ) between the arrow and the s-axis represents the principle section plane angle of the polarizer. Another arm on the right holds another Glan–Taylor prism (the analyzer) and a photomultiplier tube to detect the transmitted light exiting the analyzer.

4.2.1.1 Birefringence of calcite

The $\lambda/4$ plate and those two prism polarizers are made of calcite (calcium carbonate crystal) which has birefringent or birefractive properties. Passing through calcite crystal, an un-polarized light beam will be separated into two linear polarized beams. One following Snell's law is an ordinary ray (o-ray), and another one violating Snell's law is an extraordinary ray (e-ray). The refractive indices of the o-ray and the e-ray are $n_o = 1.6557$ and $n_e = 1.4849$ respectively for He–Ne laser (632.8 nm). If the incident plane of light is coincident with a principle section of the crystal, the polarization plane of the o-ray is perpendicular to the principle section, and the polarization plane of the e-ray is parallel with the principle section. The optical anisotropic crystal is easy to be cleaved into rhombohedra, and each of them has six equal parallelogram surfaces. The parallelogram or rhombus has two sharper angles (78°08′) and two blunt angles (101°52′). There are two obtuse spatial corners in the rhombohedron. Each of them is an intersection of three blunt angles contributed by three rhombi. A virtual line through an obtuse spatial corner of the rhombohedron is called the 'optical axis' of calcite. In geometry, a spatial angle of the optical axis to each of those three rhombi surfaces is the same. In optics, an un-polarized ray propagating along the optical axis will not be separated into two rays with two different speeds. There are no birefractive phenomena for an incident light along the axis. The plane with the optical axis perpendicular to two opposed faces of the rhombohedron is the principle section of the calcite crystal. There are three principle sections at any point inside the crystal corresponding to its three pairs of opposed faces.

A quarter-wave plate is made by cutting a calcite crystal into a slab with its optical axis parallel with its two bigger surfaces. The red dashed line on the left surface of the slab shown on the top-left corner of figure 4.2 is the optical axis of the calcite plate. After the light beam perpendicularly passes through the $\lambda/4$ plate, it splits into an *o*-ray and an *e*-ray traveling in the same direction. A virtual vector parallel with the optical axis is called '*e*-axis' of 'fast axis', because the *e*-ray travels faster than the *o*-ray inside the crystal. Another virtual vector perpendicular to the fast axis is named the '*o*-axis' or 'slow axis'. The least phase difference between the two rays exiting the slab is $\frac{2\pi}{\lambda}(n_o - n_e)d = \frac{\pi}{2}$, so the minimum thickness of the slab is 926.23 nm. It is too thin to polish and to hold the thin crystal. A high order wave plate, for example a $1000\frac{1}{4}\lambda$ calcite plate is practically thick enough (3.706 mm) for handling, and it works in exactly the same way as a quarter-wave plate at the specific wavelength. Passing through a $\lambda/4$ plate, the linear polarized light is converted into an elliptical polarized light if the angle between the principle section and the incident plane is not equal to 0° or 90°.

The polarizer and the analyzer in the system are two Glan–Taylor prisms [16]. Each of them is made of two right-angle calcite prisms separated by an air gap, as shown by the illustration on the top-right corner of figure 4.2. The optical axes (the dashed red line zz') of the two prisms are in the same direction. The incident angle of the light on the gap interface of the Glan–Taylor prisms is equal to the sharp angle of the right-angle prisms (38°). From the refractive indices of the *o*-ray and the *e*-ray, we know their critical angles for total reflection on the air gap interface are 37.155° and 42.334° respectively. The *o*-ray or the *s*-polarized wave will be totally reflected to a blackened side of the polarizer to reduce unwanted reflection. Only the *e*-ray or the *p*-polarized light will emit out of the polarizer in the direction of the incident light. By rotating the principle section of the Glan–Taylor prism, we can change the polarization plane of the *e*-ray in space.

4.2.1.2 Alignments of optical components

We need to align all the optical components in their positions and integrate them to make a null ellipsometer. A BK7 glass slab could be used as a sample for the alignments. Without the polarizer and the compensator in the left arm, we set the incident angle of the laser to be 56.6° (the Brewster's angle for the air/BK7 interface). The reflected light from the sample is polarized to be a TE wave. On the right side *s*–*p* Cartesian system of figure 4.2, the length of the arrow A_r represents the amplitude of the electric field of the reflected light in the principle section of the analyzer. The angle between A_r and the *s*-axis is the amplitude angle (Ψ) of ellipsometry. While adjusting the analyzer to find a position for the photomultiplier tube to receive the least transmitted light, the photomultiplier tube's signal changes in a profile according to Malus's law: $\cos^2(\Psi)$ for the reflected TE wave. With $\Psi = 0°$, the light detector gives the maximum signal, but the photomultiplier tube and our eyes are more sensitive to low light intensity, so we have to turn the analyzer 90° for null reflections. A direct reading angle of the analyzer at null conditions is defined as $\psi = \Psi + 90°$. By adjusting the principle section of the analyzer at a position where the transmitted light of the analyzer is extinguished, we mark the analyzer position on its scale to $\psi = 90°$. Then the analyzer is rotated 90° clockwise to a position of $\psi = 0°$, so that the principle section of the analyzer is perpendicular to the incident plane, and parallel with the polarization plane of the reflected light.

The polarizer is then reinstalled in the left arm, and the two arms are rotated in a 'straight-through' position. The transmitted light from the polarizer directly goes into the right arm. The signal of the photomultiplier tube varies with φ according to the function $\cos^2(\varphi)$, where φ is the angle between the arrow A and the s-axis in the left s–p Cartesian system. By rotating of the polarizer to find the null position for the photomultiplier tube, the polarizer scale in its holder is marked $\varphi = 90°$. Then the polarizer is rotated 90° clockwise to marking its position for $\varphi = 0°$. The polarizer alignment is done.

The compensator is then installed into the left arm while the two arms are still in the 'straight-through' position. Only when the e-axis or the o-axis of the compensator is parallel with the arrow A in the left s–p Cartesian systems of figure 4.2, can null positions be achieved by adjusting the analyzer. Under the conditions of $\varphi = 45°$and $\psi = -45°$, we can set the e-axis of the compensator at $\vartheta = 45°$ or $\vartheta = -45°$ for null reflections. After the polarizer scale (φ), the compensator scale (ϑ) and the analyzer scale (ψ) are set in the positions of their Cartesian systems, we can rotate the two arms to set the incident angle (θ) of the laser, and the null ellipsometer is ready to be used.

4.2.2 Measuring Φ and Ψ with a null ellipsometer

In figure 4.2, the compensator angle is set at $\vartheta = 45°$. The rotatable polarizer's position φ could be any number between −180° to 180°. Exiting from the compensator to the reflecting point on the sample, the transmitted light becomes an elliptical polarized light with two components (see the middle double Cartesian systems in figure 4.2)

$$\begin{cases} E_e = A\cos\left(\varphi - \dfrac{\pi}{4}\right)e^{i(kr-\omega t)} \\ E_o = A\sin\left(\varphi - \dfrac{\pi}{4}\right)e^{i\left(kr-\omega t+\frac{\pi}{2}\right)} \end{cases} \tag{4.3}$$

In the above expressions, A is the electric field amplitude of a linear polarized light wave exiting from the polarizer, $k = n_0\frac{2\pi}{\lambda}$ is the wavenumber vector length of the light, and the extra $\frac{\pi}{2}$ comes from the refractive indices difference of the calcite compensator to the two rays. The ellipse in figure 4.2 was drawn according to equation (4.3). It is the trajectory of the electric field vector of the elliptical polarized light emerging from the $\lambda/4$ plate. The length of the ellipse's semi-major axis is $A\cos(\varphi - \frac{\pi}{4})$, and the length of its semi-minor axis is $A\sin(\varphi - \frac{\pi}{4})$. An outside rectangular ABGH whose sides are tangent with the ellipse has two diagonals: AG and BH in the figure. The sharper angle between the two diagonals is labeled ξ in the figure. We can directly measure ξ and use it for Φ measurements.

When the elliptical polarized light is reflected by the sample, we have to construct a TM wave and a TE wave from the o-ray and e-ray in the double coordinate system by $E_{si} = E_e\cos(\frac{\pi}{4}) - E_o\cos(\frac{\pi}{4})$ and $E_{pi} = E_e\sin(\frac{\pi}{4}) + E_o\sin(\frac{\pi}{4})$, because we know the TM and TE wave reflectivity. They are the two components of the incident elliptical polarized light just before they are reflected

$$\begin{cases} E_{pi} = \dfrac{A}{\sqrt{2}} e^{j\left(kr - \omega t + \varphi - \frac{\pi}{4}\right)} \\ E_{si} = \dfrac{A}{\sqrt{2}} e^{j\left(kr - \omega t - \varphi + \frac{\pi}{4}\right)} \end{cases} \tag{4.4}$$

Their amplitudes are the same, but the phase difference between the TM and TE waves is $(\phi_p - \phi_s)_i = 2(\varphi - \frac{\pi}{4})$. If $\frac{\pi}{4} < \varphi < \frac{\pi}{2}$ as shown in figure 4.3(A), we have $\xi = 2(\varphi - \frac{\pi}{4})$. Or if $0 < \varphi < \frac{\pi}{4}$, we have $\xi = 2(\frac{\pi}{4} - \varphi)$. Since φ could be set manually at any angle from −180° to 180°, it is possible to adjust $(\phi_p - \phi_s)_i$ to compensate the

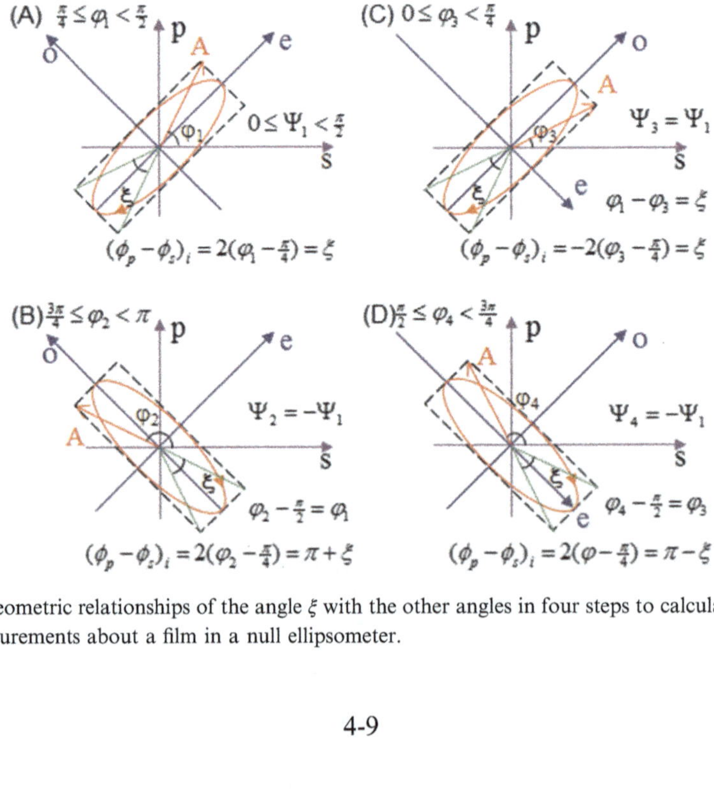

Figure 4.3. Geometric relationships of the angle ξ with the other angles in four steps to calculate Ψ and Φ for accurate measurements about a film in a null ellipsometer.

reflected light to be a linear polarized light. If the transmitted light of the analyzer can be extinguished by alternately adjusting the polarizer and the analyzer, the linear polarization status of the reflected light could be confirmed. The ellipsometry ratio ρ of the optical system shown in figure 4.2 is:

$$\rho = \tan(\Psi)\, e^{j\Phi} = \frac{r_p}{r_s} = \frac{|E_{pr}|}{|E_{sr}|} e^{j[(\phi_p - \phi_s)_r - (\phi_p - \phi_s)_i]} \tag{4.5}$$

In the above formula, ellipsometry ratio p is an unknown constant for a given sample under investigation and a monochromatic light at an incident angle θ. Its value is independent of the angles of the polarizer, φ, and the compensator, ϑ. By using the angle ξ of the ellipse, we can accurately measure Φ and Ψ with the following four steps (figure 4.3).

Step (A): Set $\vartheta = 45°$ and if we can take measurements with $0° \leqslant \Psi_1 < 90°$ and $45° \leqslant \varphi_1 < 90°$ in the null condition shown in figure 4.3(A), the reflected light is a linear polarized wave with its vector A_r vibrating in quadrant I and III of the right s–p Cartesian system in figure 4.2: $(\phi_p - \phi_s)_r = 0$. From directly reading φ_1 and ψ_1, we obtain Ψ and Φ

$$\Psi_1 = \psi_1 - \frac{\pi}{2} = \arctan\left(\frac{|E_{pr}|}{|E_{sr}|}\right) \text{ and } (\phi_p - \phi_s)_i = 2\left(\varphi_1 - \frac{\pi}{4}\right) = \xi = -\Phi_1.$$

Our measurements could be stopped now. But no instrument is perfect; there are always some errors in the measurements. If more accurate data are needed, we should do more.

Step (B): Still keeping $\vartheta = 45°$, take one more measurement in a different region for the amplitude angle: $-90° \leqslant \Psi_2 < 0°$ and $135° \leqslant \varphi_2 < 180°$ shown in figure 4.3(B). The reflected light is still a linear polarized light, but the vector A_r is vibrating in quadrant II and IV of the right s–p Cartesian system in figure 4.2. For the sample under investigation by a monochromatic light at the incident angle θ, the ellipsometry ratio ρ is a constant:

$$\tan(\Psi_1)\, e^{j\Phi_1} = \tan(\Psi_2)\, e^{j\Phi_2} = \tan(-\Psi_2)\, e^{j[\Phi_2 + \pi]}$$

From the above equation, we have $\Psi_1 = -\Psi_2$ and $\Phi_1 = \Phi_2 + \pi$. Then by taking $\Phi_1 = -2(\varphi_1 - \frac{\pi}{4})$ and $\Phi_2 = -2(\varphi_2 - \frac{\pi}{4})$ into $\Phi_1 = \Phi_2 + \pi$, it is easy to find $\varphi_2 - \varphi_1 = \frac{\pi}{2}$.

By manipulating the data measured in the above two steps, we could get more accurate data shown below. The accuracy of ellipsometry can be improved even more with two further steps

$$\overline{\Psi} = \frac{1}{2}(\Psi_1 + |\Psi_2|),\ \overline{\varphi}_1 = \frac{1}{2}\left[\varphi_1 + \left(\varphi_2 - \frac{\pi}{2}\right)\right], \text{ and } \overline{\Phi} = -2\left(\overline{\varphi}_1 - \frac{\pi}{2}\right)$$

Step(C): Rotate the compensator to set $\vartheta = -45°$ and take measurements from the same sample for $0° \leqslant \Psi_1 < 90°$ and $0° \leqslant \varphi_1 < 45°$ shown in figure 4.3(C). Below are the two components of the incident elliptical polarized light just before it is reflected by the sample

$$\begin{cases} E_{pi} = \dfrac{A}{\sqrt{2}} e^{j\left(kr - \omega t - \varphi_3 + \frac{3\pi}{4}\right)} \\ E_{si} = \dfrac{A}{\sqrt{2}} e^{j\left(kr - \omega t + \varphi_3 + \frac{\pi}{4}\right)} \end{cases} \tag{4.6}$$

Their amplitudes are also the same, but the phase difference between the TM wave and the TE wave is $(\phi_p - \phi_s)_i = -2(\varphi_3 - \frac{\pi}{4})$. The reflected light is also a linear polarized light with its electric field vector A_r vibrating in quadrant I and III of the right *s–p* Cartesian system of figure 4.2. The following equations are based on the constant of the ellipsometry ratio ρ

$$\tan(\Psi_1)\, e^{j\Phi_1} = \tan(\Psi_1)\, e^{-j2(\varphi_1 - \frac{\pi}{4})} = \tan(\Psi_3)\, e^{j2(\varphi_3 - \frac{\pi}{4})} = \tan(\Psi_3)\, e^{j\Phi_3}$$

From the above equations, it is easy to find $\Psi_3 = \Psi_1$ and $\varphi_1 + \varphi_3 = \dfrac{\pi}{2}$. The polarizer's principle section is rotated to a position so that the incident linear polarized light's electric field vector is vibrating along another diagonal of the outside rectangular ABGH shown in figure 4.2. According to geometric structures in figure 4.3(C), the angle ξ can be calculated from φ_1 and φ_3 by: $\varphi_1 - \varphi_3 = \xi$.

Step (D): Still keeping $\vartheta = -45°$, more measurements are done for $-90° \leqslant \Psi_4 < 0°$ and $90 \leqslant \varphi_4 < 135°$ shown in figure 4.3(D). The reflected light is also a linear polarized light with its electric field in quadrant II and IV of the right *s–p* Cartesian system of figure 4.2. Because the ellipsometry ratio ρ should be the same complex number, the following results are obtained: $\varphi_4 - \varphi_3 = \dfrac{\pi}{2}$ and $\Psi_4 = -\Psi_3$.

By manipulating all data obtained in the above four steps, we can obtain the average phase angle $\overline{\Phi}$ and average amplitude angle $\overline{\Psi}$. From this, we have more accurate information about any sample under investigation

$$\overline{\Psi} = \frac{1}{4}(\Psi_1 + |\Psi_2| + \Psi_3 + |\Psi_4|) \tag{4.7}$$

$$\overline{\xi} = \left| \frac{1}{2}\left[\varphi_1 + \left(\varphi_2 - \frac{\pi}{2}\right)\right] - \frac{1}{2}\left[\varphi_3 + \left(\varphi_4 - \frac{\pi}{2}\right)\right] \right| \tag{4.8}$$

In the first step of the above manipulations, we start from $45° \leqslant \varphi_1 < 90°$, so it is easy to find $\overline{\Phi} = -\overline{\xi}$ from the geometric structures shown in figure 4.3. It is also possible to do this for measurements of a sample starting with $0° \leqslant \varphi_1 < 45°$. In this case, we can find $\overline{\Phi} = \overline{\xi}$ with the same manipulations

$$\overline{\Phi} = \begin{cases} \overline{\xi} & 0 \leqslant \overline{\varphi}_1 < \dfrac{\pi}{4} \\ -\overline{\xi} & \dfrac{\pi}{4} \leqslant \overline{\varphi}_1 < \dfrac{\pi}{2} \end{cases} \tag{4.9}$$

The data [17] shown in table 4.1 is about a SiO_2 film over silicon and an Al_2O_3 film over silicon. The incident angle of a He–Ne laser was 70° in a null ellipsometer used for the measurements. The silicon refractive index for the He–Ne laser directly measured by the null ellipsometer was $n_2 = 3.95 + j0.018$. For the silicon dioxide film, calculations based on formulae (4.7)–(4.9) give the following information about the SiO_2 film over silicon: $\overline{\varphi}_1 = 85.1°$, $\xi = 82.6°$, $\Phi = -82.6°$ and $\Psi = 31.4°$. Information about the Al_2O_3 film on silicon is also obtained: $\overline{\varphi}_1 = 18.1°$, $\xi = 54.5°$, $\Phi = 54.5°$ and $\Psi = 44.0°$.

With the phase angle and amplitude angle information obtained from measurements, we can search a table to find the thickness and the refractive index of the sample under investigation. The Mathcad program shown in figure 4.4 can be used to automatically search and find the sample's parameters (n, d). The parameters of the SiO_2 film were found to be: $n_1 = 1.472$ and $d_{min} = 75.250$ nm. Under optical parameters in the second line of the program, the long line with an arrow on the left could be expanded by double-clicking to show the function $\rho(n, d)$. The function is the same as the function defined in the program shown in figure 4.1, but has two variables: the sample refractive index and thickness. Then $\Phi(n, d)$ and $\Psi(n, d)$ are defined as usual and their units are in degrees. Two solid curves in an x–y plot of the program are the outputs of $\Phi(n, d)$ and $\Psi(n, d)$ respectively, and the two dashed lines come from $\Phi0 = -82.6°$ and $\Psi0 = 31.4°$ of the sample measured by the null ellipsometer. Both Ψ and Φ change with the film thickness with a spatial period:

$$\tau = \frac{\lambda}{2\sqrt{n_1^2 - n_0^2 \sin^2(\theta)}} \tag{4.10}$$

In the case $n_1 = 1.472$, the program showed $\tau = 279.3$ nm. The program only finds the minimum thickness of the sample. From the x–y plot of the program, we can see that a SiO_2 film with a thickness of ($\tau + d_{min}$) also gives $\Phi0 = -82.6°$ and $\Psi0 = 31.4°$. This program is not smart. It just searches in a 2D space within ($1.45 \leqslant n \leqslant 1.5$) and ($0 \leqslant d \leqslant \tau$) to find an optimum point (n, d) to minimize $\sqrt{(\Phi - \Phi0)^2 + (\Psi - \Psi0)^2}$. After searching the 2D space, the sub-program $n(\Phi0,\Psi0)$ reports the sample film parameters on the bottom of the program. The parameters of the Al_2O_3 film were

Table 4.1. Measurements of Φ and Ψ. Data from [17] with permission.

SiO_2 Film				Al_2O_3 Film			
λ/4 Plate	Polarizer	Analyzer		λ/4 Plate	Polarizer	Analyzer	
ϑ(deg)	φ(deg)	ψ(deg)	Ψ = ψ−90	ϑ(deg)	φ(deg)	ψ(deg)	Ψ = ψ−90
45	85.0	122.4	32.4	45	17.8	132.9	42.9
	175.2	59.2	−30.6		108.4	44.4	−45.6
−45	2.6	121.2	31.2	−45	72.6	133.4	43.4
	92.4	58.8	−31.2		162.6	45.9	−44.1

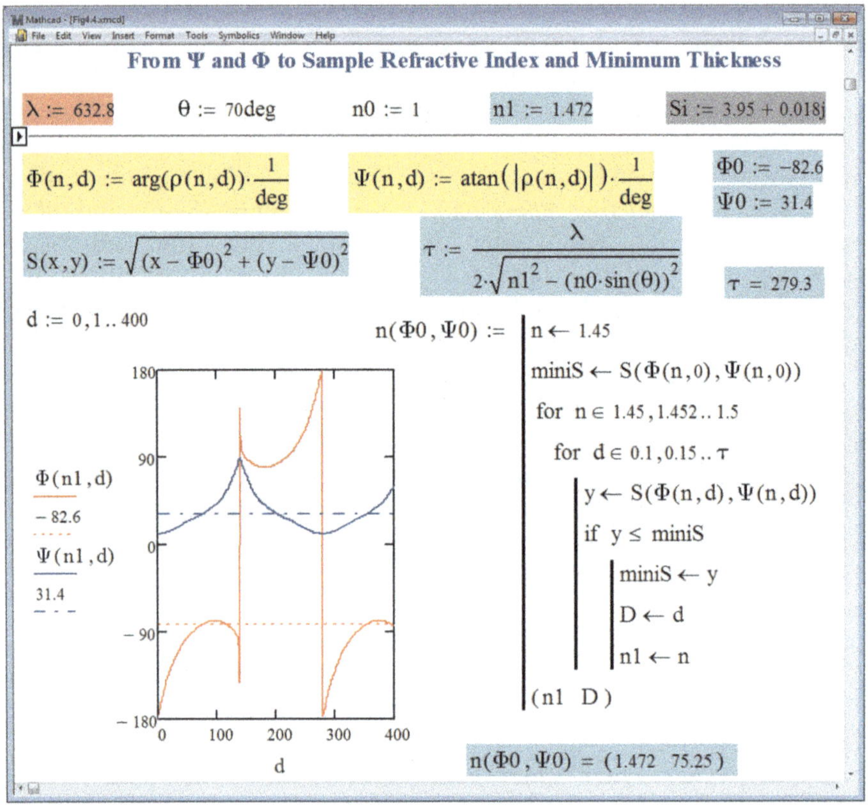

Figure 4.4. Mathcad program to extract the refractive index (1.472) and thickness (75.420 nm) of a SiO_2 film over silicon with $\Phi 0 = -82.6°$ and $\Psi 0 = 31.4°$ measured by a null ellipsometer. Available online at http://iopscience.iop.org/book/978-0-7503-2384-0.

also found to be: $n_1 = 1.755$ and $d_{min} = 127.81$ nm. The Mathcad program shown in figure 4.4 can only be used for any sample whose refractive indices are real numbers. For thin metal films with complex refractive indices, we have to use a smarter way to extract film information. There are automatic ellipsometers, such as HORIBA Jobin Yvon spectroscopic ellipsometers, and they can be used to measure Φ and Ψ quickly at different wavelengths from UV to infrared light.

4.3 SPREP: a hybridization of SPR and ellipsometry

In 1997, Rekveld [18] first applied total internal reflection ellipsometry to monitor protein adsorption on TiO_2 films. Combining the SPR technique with ellipsometry, Westphal and Bornmann [19] in 2002 invented a hybridized optical technique with a great improvement of sensitivity and reliability for their instrument. They named it 'surface plasmon enhanced ellipsometry' (SPEE). But some pioneers [20] in the new technique preferred to call it 'total internal reflection ellipsometry' (TIRE). In this book, 'SPR ellipsometry' (SPREP) is used for the name of the hybridization of the two optical techniques.

We can use total internal reflection ellipsometry to measure refractive indices of glass prisms. A semi-cylinder glass prism is horizontally put on the sample stage of a null ellipsometer shown in figure 4.5. A laser beam is perpendicular to the cylinder surface of the prism. The reflection point inside of the prism is the center of the diameter of the half cylinder. With two variables n_0 and n_2 to represent the refractive indices of the glass and the air or liquid respectively, we are able to calculate the critical angle by $\theta_c = \arcsin(\frac{n_2}{n_0})$ for total internal reflection on the interface if there is not a thin film in the system.

If $\theta > \theta_c$, the refractive angle is a complex number: $\cos(\theta_2) = j\sqrt{\frac{\sin^2(\theta)}{\sin^2(\theta_c)} - 1}$, and then we obtain explicit expressions of the Fresnel coefficients for the TM wave and the TE wave on the glass/air interface in total reflection

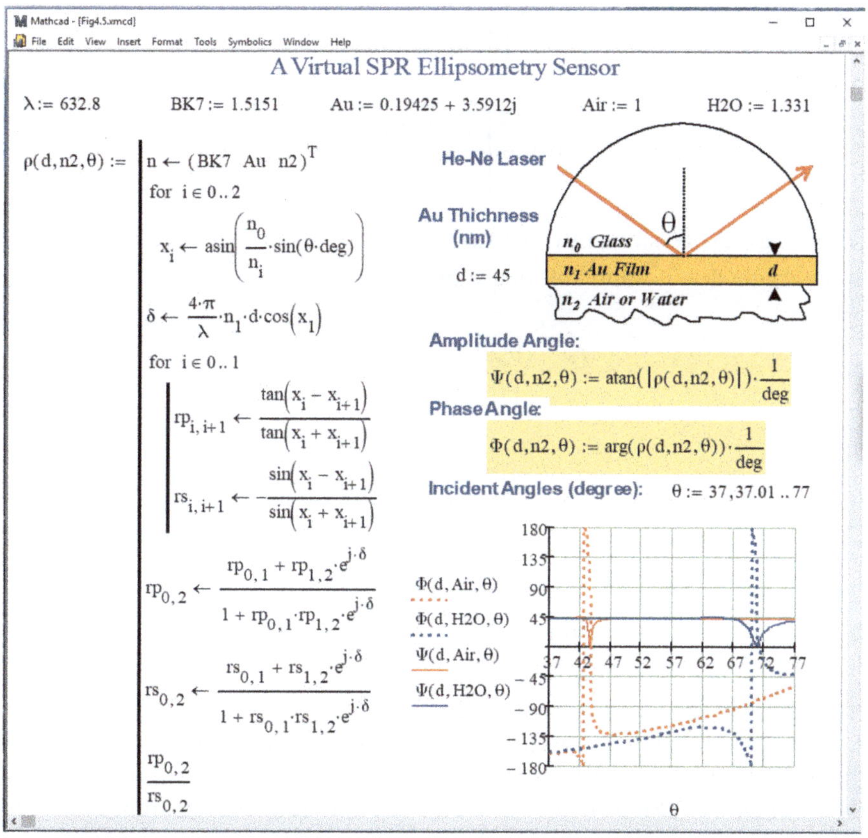

Figure 4.5. Mathcad program for numerical simulations of a virtual SPREP sensor. In the simulations, a BK7 glass prism is used to support a thin gold film. The optimum gold film thickness is 45 nm. Available online at http://iopscience.iop.org/book/978-0-7503-2384-0.

$$r_p = \frac{n_2 \cos(\theta) - jn_0\sqrt{\dfrac{\sin^2(\theta)}{sin^2(\theta_c)} - 1}}{n_2 \cos(\theta) + jn_0\sqrt{\dfrac{\sin^2(\theta)}{sin^2(\theta_c)} - 1}} = e^{j\phi_p} \tag{4.11}$$

$$r_s = \frac{n_0 \cos(\theta) - jn_2\sqrt{\dfrac{\sin^2(\theta)}{sin^2(\theta_c)} - 1}}{n_0 \cos(\theta) + jn_2\sqrt{\dfrac{\sin^2(\theta)}{sin^2(\theta_c)} - 1}} = e^{j\phi_s} \tag{4.12}$$

$$\phi_p = -2\arctan\left[\frac{n_0^2}{n_2^2}\frac{\sqrt{n_0^2 \sin^2(\theta) - n_2^2}}{n_0 \cos(\theta)}\right] \tag{4.13}$$

$$\phi_s = -2\arctan\left[\frac{\sqrt{n_0^2 \sin^2(\theta) - n_2^2}}{n_0 \cos(\theta)}\right] \tag{4.14}$$

By taking these expressions into the ratio of the two Fresnel coefficients, we find the amplitude angle $\Psi = \dfrac{\pi}{4}$ and the phase angle Φ

$$\rho = \frac{r_p}{r_s} = \frac{|r_p|e^{j\phi_p}}{|r_s|e^{j\phi_s}} = \tan\left(\frac{\pi}{4}\right)e^{j\Phi} \tag{4.15}$$

$$\Phi = \phi_p - \phi_s = -2 \arctan\left(\frac{\sqrt{\sin^2(\theta) - \sin^2(\theta_c)}}{\sin(\theta)\tan(\theta)}\right) \tag{4.16}$$

$$\frac{n_2}{n_0} = \sin(\theta)\sqrt{1 - \tan^2\left(\frac{\Phi}{2}\right)\tan^2(\theta)} \tag{4.17}$$

In the Mathcad program shown in figure 4.5, an amplitude angle $\Psi(d, n2, \theta)$ and a phase angle $\Phi(d, n2, \theta)$ functions are defined. If we set the gold film thickness as $d = 0$ and the incident angle $\theta = 70°$, we can find $\Psi(0, \text{Air}, 70) = 45°$ and $\Phi\ (0, \text{Air}, 70) = -29.049°$. Then by taking the θ value and Φ value into the formula (4.17), we find the refractive indices ratio between the air and the BK7 glass: 1/1.5151.

The Mathcad program presented in figure 4.5 is very similar to the Mathcad program shown in figure 4.1, but here the function $\rho(d, n2, \theta)$ has three local variables: the thickness of the gold film, the refractive index of the low density medium and the incident angle. Using the first line of the ρ function, we set BK7 glass for the prism to support the gold film. An x–y plot on the lower-right corner of the figure is the output of the program. The two dashed curves are the phase angle (Φ) spectra and two solid curves are the amplitude angle (Ψ) spectra. The labels on the left side of vertical axis of the x–y plot indicate that the red and blue curves are the results of using air and water as the low density medium in the system respectively. The Ψ spectra are very similar to the traditional SPR curves, and their resonance angle θ_{SPR} positions are the same as the SPR spectra. The optimum gold film thickness is 45 nm. If the gold film is thinner, its SPR adsorption valley will be shallow. For thicker gold films, the spectra are sharper for both Ψ and Φ spectra. Because two polarized light waves are used for mutual reference in SPREP, measurements about the relationship between Ψ and θ are more reliable than traditional SPR spectra measurements. In a narrow region around the θ_{SPR} in the SPREP system, the phase spectra have a significant change. The change could be from $-180°$ to $180°$. That is the reason why SPREP sensors are very sensitive for biosensing applications.

4.4 Using the Levenberg–Marquardt algorithm to interpret SPREP spectra

After using an SPR ellipsometer to obtain SPREP spectra of a gold film, we need to extract information about the thickness and the refractive index of the sample under investigation. Taking $(N + 1)$ points on a Ψ spectrum: $\{(\Theta_i, y_i), \ i = 0,1,\ldots, N\}$, we can use the Levenberg–Marquardt algorithm [21, 22] (LMA) to write a Mathcad program for a least squares curve fitting of the SPREP Ψ spectrum, and we will then have a reliable program to find gold film parameters from a group of guessed numbers. As a modification of the Gauss–Newton algorithm, LMA is more robust. Even those initial numbers are very different from true values of parameters for a theoretical model function, we can still use LMA to find a good solution to our problems. A theoretical function $y = f(t, X)$ is used to model a set of observed data ($N + 1$ points). This function has $(M + 1)$ parameters to be determined and they are organized in a $(M + 1) \times 1$ matrix as $X = [\chi_0, \chi_1, \ldots, \chi_M]^T$, where N and M are integers with $N \geqslant M$. A function $S(X)$ is constructed and it is a half of the sum of the squared residuals between the model values $f_i = f(t_i, X)$ and the observed values y_i at all observing points ($i = 0, 1\ldots N$)

$$S(X) = \frac{1}{2}\sum_{i=0}^{N}[f_i - y_i]^2 = \frac{1}{2}R^T \cdot R \tag{4.18}$$

where $R = [r_0, r_1, \ldots r_N]^T = [F(X) - Y]$, $F(X) = [f(t_0, X), f(t_1, X), \ldots f(t_N, X)]^T$, and $Y = [y_0, y_1, \ldots y_N]^T$ are three $(N + 1) \times 1$ matrixes respectively. The optimal parameter matrix X that minimizes $S(X)$ comes from a condition by which the gradient of $S(X)$ to X is equal to zero [23]

$$\nabla S(X) = J^T \cdot R = 0 \tag{4.19}$$

where J is the same Jacobian matrix as in section 1.2.1 with $J_{ij} = \frac{\partial f_i}{\partial \chi_j}$, ($i = 0, 1, \ldots, N$, and $j = 0, 1, \ldots, M$). Usually, it is not easy to solve $\nabla S(X) = 0$, but we can use LMA to obtain an optimal parameter matrix from an initial X_0 by iterations

$$X_{k+1} = X_k - [J^T \cdot J + \upsilon I]^{-1} \cdot J^T \cdot R \tag{4.20}$$

Comparing with the formula (1.19) for the Gauss–Newton algorithm, we see one more matrix in the above formula: υI, where I is a $(M + 1) \times (M + 1)$ identity matrix, and υ is the damping factor. The damping factor will be adjusted automatically during iteration according to the value of $S(X)$. If $S(X_{k+1})$ is smaller than $S(X_k)$, then the iteration is in the correct direction, and 0.1υ will be used to replace υ for the next interaction to accelerate the iteration processes. Otherwise, the iteration is in the wrong direction, and 10υ will be used to replace υ to change the iteration direction.

The Mathcad program based on LMA shown in figure 4.6 is used to extract gold film information from amplitude angles obtained from a system presented in the top-right corner of the figure. A gold film is directly deposited on the base of a semi-cylinder prism made of BK7 glass, and the gold film is contacted with water. In the second line of the program, parameters of the system are assigned to variables: λ for the wavelength of the light source, BK7 for the prism refractive index, and H2O for water refractive index. In the same line, the damping parameter υ is set to be 1000; the parameter Δ is used as a stop sign for iterations. A vector X is used to store the guessed initial parameters for the real part (η), the imaginary part (κ) of a gold film's refractive index, and the gold film's thickness (d) respectively. These three parameters would be optimized at the end of the program. Since there are three parameters to be optimized, a 3×3 identity matrix is assigned to a variable 'I' which will be used in iterations.

The fourth line on the left of the program defines the theory function for the amplitude angle Ψ with four variables: η, k, d and θ. The first line in the function combines the first two variables into the gold refractive index: Au $\leftarrow \eta + j \cdot k$. The remaining lines in the function are almost the same as those in figure 4.5. The last line of the function returns an amplitude angle value Ψ in degrees when the function is called. All notes in blue on the interface of the program are text regions. Having no experimental data, I have to use the theory function $\Psi(\eta, k, d, \theta)$ to create a set of 'observed' data around the surface plasmon resonance angle (~71°) shown in the lower-right corner of the program with pink colored parameters $\eta = 0.194\ 25$, $k = 3.5912$ and $d = 42$ nm above an x–y plot and at incident angles (65, 68, 71, 74, 77). The amplitude angles are obtained by calling $\Psi(\eta, k, d, \theta)$ to get: (44.1, 39.0, 10.6, 33.7, 39.2) respectively showing at the lower-right corner of the program. Copy the two arrays and paste them to 'Incident Angles' Θ and 'Input Ψ ===>' y in the third and fourth line of the program respectively.

Suppose the set of data with five ($N = 0\ldots4$) points are experimental data coming from a system with known n_0 and n_2, but unknown n_1 and d, we define another sub-program name 'G' to manipulate these data to find the true parameters of the gold film from 'Initial Parameters' stored in a vector X. The component $X_0 = 0.5$ is the guessed real part of the gold refractive index, $X_1 = 5$ is the guessed imaginary part of the gold refractive index, and $X_2 = 200$ is the guessed thickness of the gold film. The

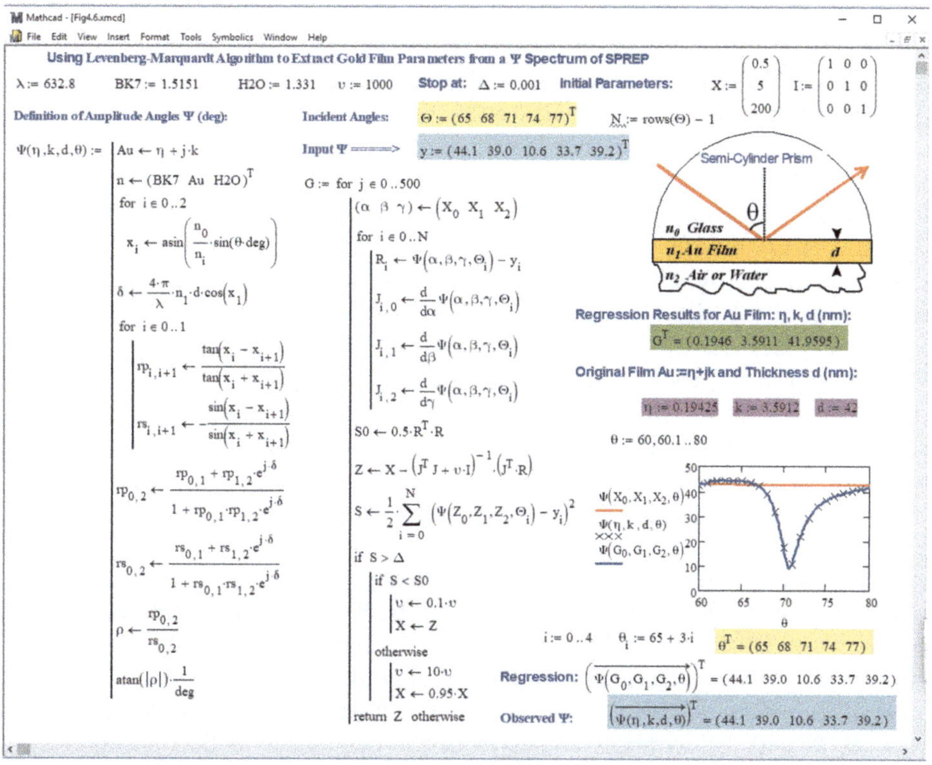

Figure 4.6. Mathcad program based on the Levenberg–Marquardt algorithm for nonlinear regressions to interpret Ψ data for gold film parameters. To use the program, five experimental data points need to be input into the program. Available online at http://iopscience.iop.org/book/978-0-7503-2384-0.

reliability of the Mathcad program can be tested by changing these three initial parameters. If the nonlinear regression program is not sensitive to the initial parameters, the program is reliable.

The sub-program $\Psi(\eta, k, d, \theta)$ on the left of the figure is the theory function for the nonlinear regression program based on the Levenberg–Marquardt algorithm. Observed data should fit with it, and the constant 'Δ' is used to quantify the fitness. The sub-program 'G' with a long vertical line has a loop for an integer variable j changing from 0 to 500. This is written according to LMA. Because of the reliability of the program, only a few iteration steps are enough to obtain optimal parameter values. The second line of the sub-program G is used to assign the guessed initial values to α, β and γ respectively. The theory value of the function is $\Psi(\alpha, \beta, \gamma, \Theta)$ in the sub-program for numeric calculations. With these initial values, a loop with an integer 'i' increasing from 0 to N calculates element values of the R vector and J matrix at each observing point by $R_i = \Psi(\alpha, \beta, \gamma, \Theta_i) - y_i$, and $J_{i,0} = \frac{d}{d\alpha}\Psi(\alpha, \beta, \gamma, \Theta_i)$, $J_{i,1} = \frac{d}{d\beta}\Psi(\alpha, \beta, \gamma, \Theta_i)$, $J_{i,2} = \frac{d}{d\gamma}\Psi(\alpha, \beta, \gamma, \Theta_i)$.

Then the sub-program G evaluates the difference ($S0$) between the theory function and observed values at all data points according to formula (4.18). After modifying the initial guessed vector X to Z according to equation (4.20), the difference (S) is evaluated again with a new vector Z. If the S value is bigger than Δ ($= 10^{-3}$), more iterations are needed to process with two options. (1) If $S < S0$ is true, the iteration is in its way, the damping factor υ is replaced with 0.1υ to accelerate the iteration process, and the vector Z is assigned to X for the next iteration. (2) Otherwise, the damping factor υ is replaced with 10υ, the original vector X is replace by $0.95X$, and the vector Z is discarded. (I added the step '$X \leftarrow 0.95X$' to modify LMA, so the Mathcad program is more reliable.) Then the sub-program will increase the integer j by one, assign the new vector X elements to α, β, and γ, and repeat the above calculations until the condition $S < \Delta$ is true. The output of the sub-program G is the final vector Z with three optimized parameters of the gold film. G^{T} under the schematic of the optical system on the right of the program is the output of the sub-program. With the transpose operation symbol 'T' to the vector G, we can write the three parameters of the vector in a line to save space on the program interface. The regression results are very close to the original gold film parameters listed under G^{T} highlighted in pink.

An x–y plot inside the program shows three Ψ spectra with the incident angle changing from 60° to 80° and a step of 0.1°. The red curve $\Psi(X_0, X_1, X_2, \theta)$ is based on the initial guessed parameters. As the initial gold thickness is 200 nm, the film acts almost like bulk gold, and Ψ is almost a constant close to 42.4°. A spectrum $\Psi(\eta, k, d, \theta)$ with '×' symbols comes from the original gold film parameters. The regression spectrum $\Psi(G_0, G_1, G_2, \theta)$ (blue) fits well with $\Psi(\eta, k, d, \theta)$.

It is reasonable to think that the Mathcad program is reliable because it uses its own data to prove itself. It creates a set of amplitude angles $\{\Psi\}$ (44.1, 39.0, 10.6, 33.7, 39.2) at five incident angles $\{\theta\}$ (65, 68, 71, 74, 77) based on a 42 nm gold film with refractive index $n1 = 0.194\,25 + j3.5912$. Changing the second line of the $\Psi(\eta, k, d, \theta)$ function from '$n \leftarrow$ (BK7, Au, H2O)' to '$n \leftarrow$ (H2O, Au, BK7), we have a virtual traditional ellipsometer, in which the same gold film is deposited on the BK7 glass in water. All the data at the five points fit exactly, but the gold parameters from the regression are very different from the original gold parameters. By changing the incident angle region from 1° to 90°, we can see that the traditional ellipsometry Ψ values were around 42°. Even if we choose Ψ points at incident angles from 5° to 85° with a step of 20°, the program still cannot find the original gold film parameters in the traditional ellipsometer.

Changing the second line of $\Psi(\eta, k, d, \theta)$ function to '$n \leftarrow$ (BK7, Au, 1)', we have a virtual SPR ellipsometer in which the low density medium is air. The program still works well. Changing the last line of the $\Psi(\eta, k, d, \theta)$ function to $\arg(\rho) \cdot \frac{1}{\deg}$, we define a theory function for the phase angle spectrum. The program still works, but it is more sensitive to the initial parameters. In figure 4.5, we see the phase spectra are not continuous at incident angles around the surface plasmon resonance angles. This could be a reason why the program is not reliable with Φ spectra data. If we want a virtual SPR ellipsometer with Otto configuration to excite surface plasmons on an air/gold interface, the second line of the $\Psi(\eta, k, d, \theta)$ function should be replaced by '$n \leftarrow$ (BK7, 1, Au)'. In this way the program could be used to find the air gap thickness and the gold refractive index. The Mathcad program shown in figure 4.6 is also used in section 4.5 for super biosensors.

4.5 A virtual SPREP biosensor

SPR ellipsometry (SPREP) biosensors are so sensitive that we do not need to write a section about any traditional ellipsometry biosensor. Readers could modify the Mathcad program shown in figure 4.7 to build a virtual traditional null ellipsometry biosensor without a chrome film ($d1 = 0$). By setting the incident angle θ at 57°, we would have a benchmark $\gamma = 1.913$ [deg nm^{-1}] from its Φ spectra. It is really more sensitive than SPR biosensors, but its linear region is less than 5 nm. It is almost useless for biosensing applications, especially for affinity measurements.

The optical system of a virtual SPREP biosensor shown in figure 4.7 has three thin films and five media. On the base of a BK7 glass prism, there is a chrome film ($d1 = 2$ nm) and a gold film ($d2 = 10$ nm). A thin film of chrome or titanium on glass is usually used to enhance the adhesion of gold to glass. Since the SPREP biosensor uses both a TM wave and a TE wave, two Fresnel coefficients are calculated on each interface of the system in the definition of the ellipsometry's phase angle function $\Phi(n3, d3, \theta)$. The three local variables in the function are the sample refractive index $n3$, thickness $d3$, and the incident angle θ. The last line of the function returns a value of Φ in degrees when the function is called. Curves of $\Phi(n3, d3, \theta)$ in an x–y plot under the schematic of the SPREP prism system are Φ spectra. At an incident angle around the critical angle θ_c (61.5°), the phase angle of the system suddenly flips over from −180° to 180°, and then slowly decreases to 0°. Those three Φ curves construct a spindle in a region of $62° < \theta < 90°$. In the middle of the spindle ($\theta \sim 70°$), Φ values are most sensitive to the sample film. It is convenient to adjust the incident angle θ to be fixed at $\Theta = 70°$ to observe Φ changing while a sample film on the gold film is growing.

Under the SPREP spectra on the program interface, one more function $Z(n3, d3, \theta)$ is defined as a ratio between $\Delta\Phi$ and $(n3 - n4)$. Linear regions of three curves for $Z(n3, d3, \theta) \sim d3$ are searched, and the slope of a curve about a relationship between $Z(1.4, d3, \theta)$ and $d3$ is used as the benchmark of the SPREP biosensor. Linear correlation coefficients are set to be bigger than or equal to 99.9%. By setting the He–Ne laser incident angle at $\Theta = 70°$, we found the linear region of the SPREP system was up to 40 nm, and its benchmark was 7.35. The sensitivity of the SPREP system is ten times more than that of BIAcore 1000. Table 4.2 shows more simulation results.

The greatest advantage of the virtual SPREP system is that it has a tunable sensitivity. The benchmark of the system can be tuned from 6.7 to 50.8 while its linear region can be changed from 60 nm to 9 nm. To assay samples with bigger sizes such as antibodies, we can tune the SPREP biosensor to a longer linear region and a lower sensitivity. For samples with smaller molecular sizes such as insulin, we need a higher sensitivity and a shorter linear region. The laser incident angle, the

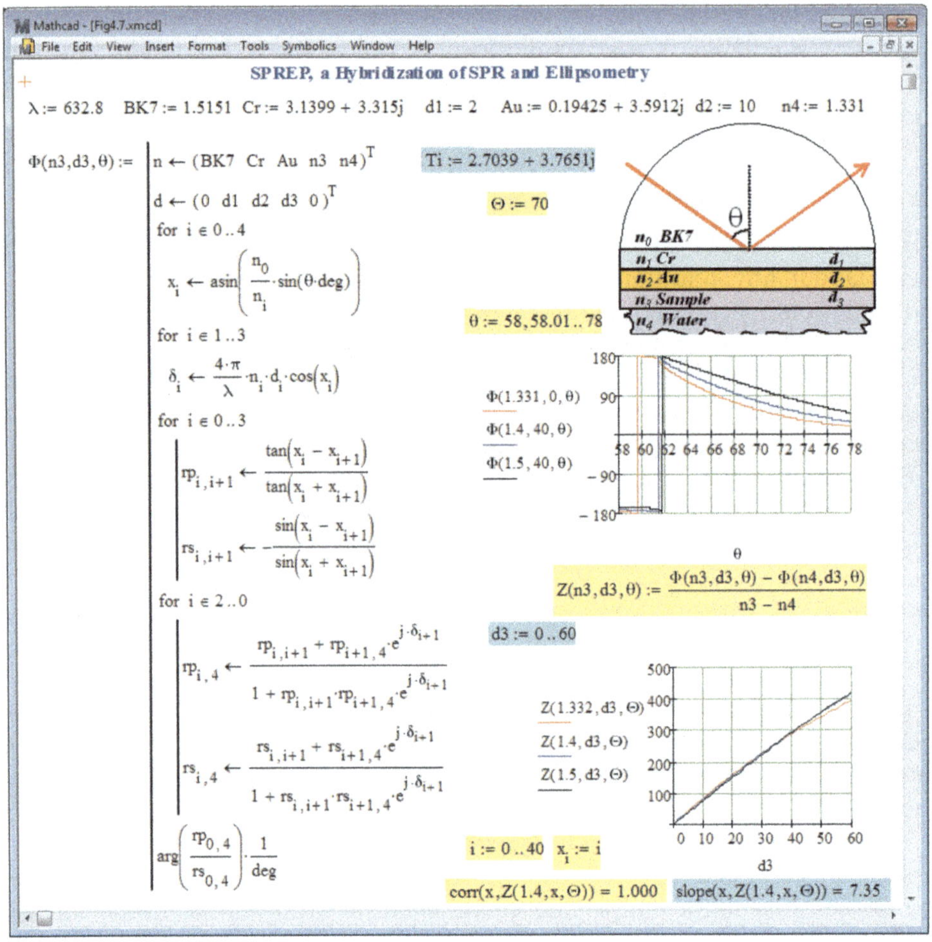

Figure 4.7. Numerical simulations of a virtual SPREP biosensor with three thin films. Under the optical schematic (not to scale) are SPREP spectra. Another *x–y* plot for a function *Z*(*n*3, *d*3, *θ*) is used to find the linear region and the benchmark (the slope) of the SPREP system. The program is available online at http://iopscience.iop.org/book/978-0-7503-2384-0.

thicknesses of gold film, chrome film or titanium film can all be used to tune the system's sensitivity.

From the table, we see a thin chrome or titanium film between the gold film and the base of the prism could enhance the system's sensitivity and extend its linear response region. The data in the second column of the table came from the system with only a gold film deposited on the base of the prism. Other data on the right of column 2 were numerical simulation results of the prism with two thin metal films. The thicknesses of the chrome or titanium films were 2, 5, 8 and 10 nm respectively. The upper half of the table is for chrome, and the other half of the table is for titanium. The data in the third column of the table indicate that a system with Cr 2 nm/Au 10 nm has a linear region up to 60 nm; but the same system without a chrome

Table 4.2. A virtual SPREP biosensor with a tunable benchmark.

Cr (nm)	0	2		5		8		10	
Au (nm)	10	10	25	10	25	10	25	10	25
γ (deg nm^{-1})	6.7	7.1	24.7	8.1	28.1	8.9	33.5	10.8	39.0
$d3_{max}$ (nm)	43	60	17	50	17	45	13	40	11
Θ (deg)	69.5	69.5	71.5	70.0	72.0	71.5	72.5	73.0	73.0
Ti (nm)	0	2		5		8		10	
Au (nm)	25	10	25	10	25	10	25	10	25
γ (deg nm^{-1})	22.9	7.3	25.5	8.7	30.4	10.1	38.6	11.5	50.8
$d3_{max}$ (nm)	17	60	18	50	17	45	13	40	9
Θ (deg)	71.2	69.5	71.5	70.0	72.0	71.5	72.5	73.0	72.6

film has a linear region up to 43 nm. From data shown in the table, we can see that titanium is better than chrome for the system. The thicker the gold film, the more sensitive the SPREP system is, but a thicker gold film significantly reduces the linear region of the system. For example, an SPREP system with Cr 10 nm/Au 25 nm has a benchmark of 39.0 and a linear region up to 11 nm; a system with Cr 10 nm/Au 10 nm has a benchmark of 10.8 and a linear region up to 40 nm.

In SPR biosensors, sharper spectra are needed for precisely searching resonance angles (θ_{SPR}) automatically by software. An SPR optical system will generate sharper spectra if it has a prism with a higher refractive index, a light source with longer wavelength, and a metal film with a smaller real part in its refractive index. Controversially, in an SPREP system, we need broad SPR spectra, so that its Φ spectra spindle shown in figure 5.2 extends as long as possible between θ_c and 90°. In this way, all Φ spectra for a growing sample film will be in the focus of an incident angle Θ (γ is less sensitive to the incident angle of observations). Why can a thin film of chrome (Cr) or titanium (Ti) underneath a gold film enhance SPREP sensitivity and linearity?

To a metal with a complex refractive index $n = \eta + j\kappa$, both the real part and the imaginary part of the refractive index are positive numbers. The smaller the value of the real part η, the sharper the SPR spectra would be generated by the metal film. Metal chrome and titanium have bigger η in their refractive indices. Combined Cr/Au or Ti/Au films could be treated as an equivalent single metal film. By adjusting the thickness of Cr or Ti film between a prism and a gold film, we could tune the refractive index of an equivalent metal film on the base of a prism to fulfill our requirements. Figure 4.8 shows the Mathcad program to find the refractive index of the equivalent single metal film.

In the second line of the program, the thicknesses of a chrome film and gold film were set to be 2 nm and 10 nm respectively, so the thickness of the equivalent single metal film was equal to 12 nm. An ellipsometry amplitude angle function ψ(Cr, Au, θ) and an ellipsometry phase angle function ϕ(Cr, Au, θ) of the SPREP system shown in the figure were defined and they were hidden in a line with an arrow to save space

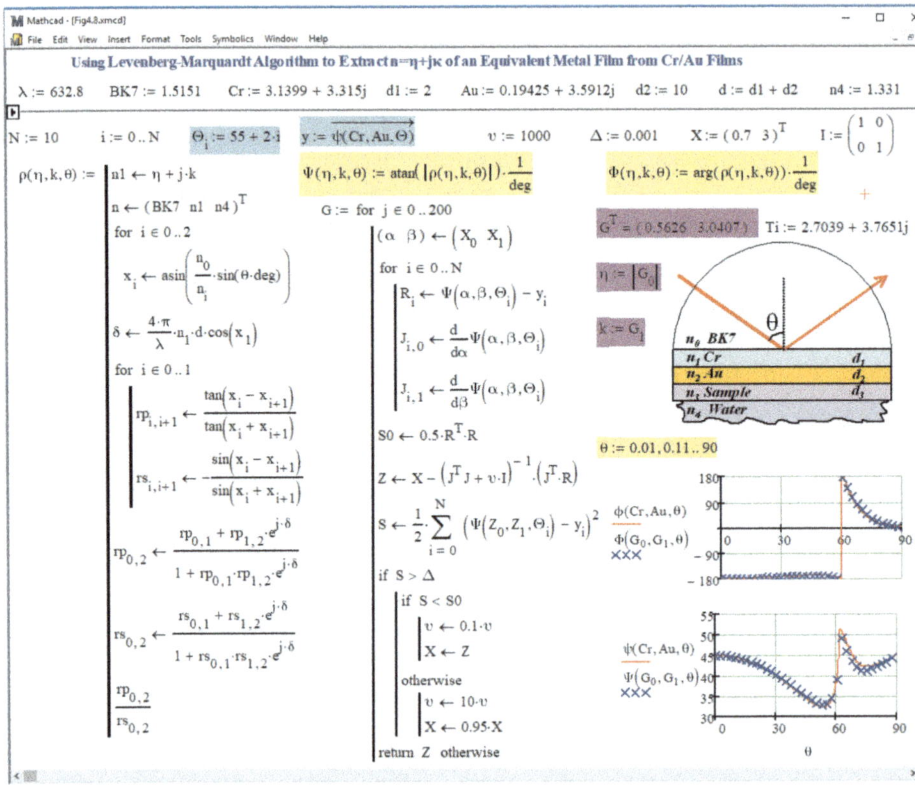

Figure 4.8. A chrome film ($n1$, $d1$) and a gold film ($n2$, $d2$) deposited on a prism could be treated as an equivalent single metal film (n, $d1 + d2$). The Levenberg–Marquardt algorithm was used to find the refractive index of the equivalent metal film. The three pink fields on the upper-left corner of the SPREP schematic were the outputs of the program. The program is available online at http://iopscience.iop.org/book/978-0-7503-2384-0.

in the program. Refractive indices of chrome and gold were local variables of these functions. Two capital Greek letters Ψ and Φ were reserved for an equivalent single metal film in the program. In the definitions of those two functions, the sample film was ignored by setting $n3 = n4$. Under the long horizontal line, 11 incident angles were chosen: $\Theta = 55, 57 \ldots 75$, and amplitude angles were calculated by calling $\psi(\mathrm{Cr}, \mathrm{Au}, \Theta)$ at those incident angles. A long arrow above the ψ function in the program is a matrix operator called 'vectorize' in Mathcad. These data were used in the following sub-program based on the Levenberg–Marquardt algorithm to find a refractive index ($n = \eta + j\kappa$) of the equivalent metal film. Two theory functions $\Psi(\eta, \kappa, \theta)$ and $\Phi(\eta, \kappa, \theta)$ were defined from $\rho(\eta, \kappa, \theta)$ for a system with an equivalent metal film deposited on the base of the prism.

The two solid red curves in two x–y plots shown on the lower-right corner of the program were from $\psi(\mathrm{Cr}, \mathrm{Au}, \theta)$ and $\phi(\mathrm{Cr}, \mathrm{Au}, \theta)$ of the system with two metal thin films, and the blue cross symbols were from $\Psi(\eta, \kappa, \theta)$ and $\Phi(\eta, \kappa, \theta)$ for an equivalent metal film system. We could see $n = 0.5626 + j3.0407$ for a combination of Cr 2 nm/Au 10 nm. Readers could change $d1$ and $d2$ manually in the Mathcad program to investigate more equivalent metal films. Here are more refractive indices of equivalent metal films: (1.8080 + j3.1998) for Cr 2 nm/Au 2 nm, (0.2136 + j3.4518) for Cr 2 nm/Au 30 nm, and (0.9385 + j2.8308) for Ti 8 nm/Au 10 nm. If you set the thickness of the chrome film or the gold film to be zero in the program, the output of the program would be the refractive index of the gold film or chrome film exactly. The combination of Cr/Au or Ti/Au could tune the properties of equivalent metal films in favor of SPREP's Φ spectra.

Phase angle functions $\phi(\text{Cr, Au}, \theta)$ and $\Phi(\eta, \kappa, \theta)$ could also be used in the sub-program of the Levenberg–Marquardt algorithm, but the regression results were not reliable. An optimum combining of titanium and gold films could be Ti 8 nm/Au 10 nm. With the He–Ne laser incident angle set at 71.5°, the benchmark of the system was 10.1. According to the formula (1.12), we have a linear relationship between $\Delta\Phi$ and a change of sample surface concentration ($\Delta\Gamma$). The linear response region of the virtual biosensor is $d3_{max} = 45$ nm, or $\Delta\Phi_{max} = 76.8°$, or $\Delta\Gamma_{max} = 42.2$ ng mm^{-2}

$$\Delta\Phi = 1.82\Delta\Gamma \tag{4.21}$$

The sensitivity of the SPREP biosensor was significantly high compared with BIAcore 1000. For every 1 ng mm^{-2} protein on a gold surface, the phase angle Φ increases 1.82°. In the case when the phase angle Φ resolution was at 0.01°, the virtual SPREP biosensor's lowest detection limit of protein on gold surface is 5.5×10^{-3} ng mm^{-2}. Human serum albumin (HSA) is a protein in human blood with a molecular weight 66.5 kDa. Responses ($\Delta\theta_{SPR}$) of SPR biosensors to a HSA film on a gold surface were about 0.5°. Responses ($\Delta\Phi$) of null ellipsometers were in a range of 1°–3°. Using a SPREP biosensor in 2007, Poksinski and Arwin [24] reported a huge response to the same sample: $\Delta\Phi = 72°$. In 2005, Nabok and his colleagues published details [25] of using a SPREP biosensor to detect an environmental toxin T2 mycotoxin and two herbicides: Simazine and Atrazine. These samples have molecular weights of 467, 202 and 216 respectively.

References

[1] Moses A J 1970 *Refractive Index of Optical Materials in the Infrared Region* (Culver City, CA: Huges Aircraft Company)

[2] Rumble J R and Rumble J 2017 *CRC Handbook of Chemistry and Physics* 98th edn (Boca Raton, FL: CRC Press)

[3] Drude P 1902 *The Theory of Optics* (New York: Dover)

[4] Rothen A 1945 The ellipsometer, and apparatus to measure thickness of thin surface films *Rev. Sci. Instrum.* **16** 26–30

[5] Rothen A 1957 Improved method to measure the thickness of thin films with a photoelectric ellipsometer *Rev. Sci. Instrum.* **28** 283–5

[6] Azzam R M, Rigby P G and Krueger J A 1977 Kinetics of protein adsorption and immunological reactions at a liquid/solid interface by ellipsometry *Phys. Med. Biol.* **22** 422–30

[7] Rothen A 1947 Immunological reactions between films of antigen and antibody molecules *J. Biol. Chem.* **168** 75–97

[8] Vladimirova L, Tsutsumanova G, Gritzkow A and Savov V 2009 Ellipsometrical and SEM study of ferritin adsorption on a gold surface *Acta Phys. Pol.* A **116** 705–7

[9] Berlinda T, Tengvallb P, Hultmana L and Arwina H 2011 Protein adsorption on thin films of carbon and carbon nitride monitored with *in situ* ellipsometry *Acta Biomater.* **7** 1369–78

[10] Nabok A and Tsargorodskaya A 2008 The method of total internal reflection ellipsometry for thin film characterisation and sensing *Thin Solid Films* **516** 8993–9001

[11] Nakanishi K, Sakiyama T and Imamura K 2001 On the adsorption of proteins on solid surfaces, a common but very complicated phenomenon *J. Biosci. Bioeng.* **91** 233–44

[12] Sarakinos J and Spyridelis J 1975 Ellipsometer data analysis of a multilayer film with a suitable computer program *Thin Solid Films* **28** 167–77

[13] Stromberg R R, Tutas D J and Passaglia E 1965 Conformation of polystyrene adsorbed at the θ-temperature *J. Phys. Chem.* B **69** 3955–63

[14] McCrackin F L, Passaglia E, Stromberg R R and Steinberg H L 1963 Measurement of the thickness and refractive index of very thin films and the optical properties of surfaces by ellipsometry *J. Res. Natl. Bur. Stand.-A. Phys. Chem.* **67A** 363–77

[15] Tompkins H G and Irene E A 2005 *Handbook of Ellipsometry* (New York: William Andrew)

[16] Archard J F and Taylor A M 1948 Improved Glan-Foucault prism *J. Sci. Instrum.* **25** 407

[17] Shao Z, Li Z, Ceng G, Fang P, Xu J, Kang R and Luo J 1974 Thickness and refractive index of thin film on Silicon surface measured by Ellipsometry *Acta Xi'an Jiatong Univ.* **2–3** 17–33

[18] Rekveld S 1997 Ellipsometric studies of protein adsorption onto hard surfaces in a flow cell *PhD Thesis* University of Twente, Enschede, Netherlands

[19] Westphal P and Bornmann A 2002 Biomolecular detection by surface plasmon enhanced ellipsometry *Sens. Actuators* B **84** 278–82

[20] Arwin H, Poksinski M and Johansen K 2004 Total internal reflection ellipsometry: principles and applications *Appl. Opt.* **43** 3028–36

[21] Levenberg K 1944 A method for the solution of certain problems in least squares *Q. Appl. Math.* **2** 164–8

[22] Marquardt D W 1963 An algorithm for least squares estimation on nonlinear parameters *J. Soc. Ind. Appl. Math.* **11** 431–41

[23] http://fourier.eng.hmc.edu/e176/lectures/NM/node36.html

[24] Poksinski M and Arwin H 2007 Total internal reflection ellipsometry: ultrahigh sensitivity for protein adsorption on metal surfaces *Opt. Lett.* **32** 1308–10

[25] Nabok A V, Tsargorodskaya A, Hassan A K and Starodub N F 2005 Total internal reflection ellipsometry and SPR detection of low molecular weight environmental toxins *Appl. Surf. Sci.* **246** 381–6

Chapter 5

Waveguide biosensors

Two new super optical biosensors are presented in this chapter. They are the development of an extinct optical biosensor named the 'resonant mirror', in which there is silica film and a titania film on the base of a glass prism. Only in a waveguide resonance condition (TM_0 mode) and internal total reflection, can light be coupled from the silica film into the titania film as a probe to detect a biological film over the titania film. By monitoring reflected light intensity at a fixed incident angle in the new biosensors, phase angle and amplitude angle (45°) of ellipsometry can be obtained easily. Also a porous silicon interferometric biosensor is mentioned.

Although SPREP biosensors are more sensitive, they are not as popular as SPR biosensors. From chapter 4, we see that the structure and operation of any SPREP biosensor is very complex. The benchmark of an SPREP system is very sensitive to its light incident angle (table 4.2). In this chapter, two more super optical biosensors are presented. They are based on a waveguide biosensing technology: resonant mirror (RM). Their structures and operations are even simpler than those of SPR biosensors. The resonant mirror (RM) biosensing technique was first reported by Cush and his colleagues [1] in 1993. They published a paper entitled 'The resonant mirror: a novel optical biosensor of direct sensing of biomolecular interactions. Part I: Principle of operation and associated instrumentation.' Commercial RM biosensors had the name 'IAsys'. They were produced by Affinity Sensors, a company in Saxon Hill, Cambridge, UK. In an RM biosensor, a silica film deposited on the base of a glass prism is used as a coupling layer, and over it is a titania film used as a resonant layer. In a total reflection, light coupled into the resonant film travels in a zigzag way just as a light propagates in an optical fiber, and is then coupled back into the prism. The coupling of a polarized light into the resonant film is dependent on waveguide modes or eigen-waves of the sensing system. Receptor molecules immobilized on the titania film can modulate the mode of the system. Without metal films in sensing chips of an RM system, an evanescent wave from internal total reflection can penetrate a few hundred nanometers into the sample layer on the

doi:10.1088/2053-2563/ab0823ch5

sensing chip. It was thought [2, 3] that RM biosensors would be more sensitive than SPR biosensors. From the following numeric simulations, we will see that RM biosensors were not sensitive. However, by a hybridization of RM with the other two optical techniques: ellipsometry and attenuated total internal reflection (ATR), we could make two super biosensors called RMEP and RMEPA respectively.

5.1 Eigenvalue equations of a slab waveguide

An optical fiber used in high speed Internet is a waveguide made of a silica core surrounded by transparent cladding with a lower refractive index. We start from a simple slab dielectric waveguide shown in figure 5.1. The slab thickness is d and its refractive index is n_2. On the bottom and the top of the slab are two plane transparent media with refractive index $n_1 < n_2$ and $n_3 < n_2$ respectively. A 3D Cartesian coordinate is set in the system with its xoy plane on the bottom of the slab, its x-axis points to the right, its z-axis points up, and its y-axis points into the page.

Although the slab dielectric waveguide is simpler than an optical fiber, we still cannot get explicit mathematical solutions about its modes or eigen-waves [4]. By replacing the dielectric media n_1 and n_3 with two metal plates, we have a waveguide for microwave transmission. In the waveguide with the simplest geometry, the slab medium is an air gap ($n_2 = 1$). The simplest waveguide is called the 'metal-walled guide'. The two metal plates work just like two parallel mirrors with a gap of d. A transverse electromagnetic (TE) wave propagates from left to right along a zigzag pathway with an incident angle θ. The frequency of the microwave could be 10 GHz ($\lambda = 3$ cm). The waveguide can be analyzed with simple mathematic manipulations. From the simplest waveguide, we can get basic ideas about possible electromagnetic waves transmitted in the waveguide. A possible electromagnetic wave is called the mode of the waveguide.

The electric field is in the y-axis direction for a TE wave traveling in the metal-walled guide. Its wavenumber vector length is $k = 2\pi/\lambda$. Along the x-axis, the length of the vector projection of $\vec{k}$ is $k_s = k\sin(\theta)$. We give k_s the name 'propagation constant'. Along the z-axis, the length of the vector projection of $\vec{k}$ is $\beta = k\cos(\theta)$ for the up and down waves respectively. The two waves superpose into a stationary wave which is a solution of a wave equation (2.23): $E_y = A_1 e^{j[k_s x + \beta z - \omega t]} + A_2 e^{j[k_s x - \beta z - \omega t]}$

In the above expression, A_1 and A_2 are unknown constants. In any metal plate, the strength of the electric field is zero. From a boundary condition $\vec{n} \times (\vec{E}_2 - \vec{E}_1) = 0$

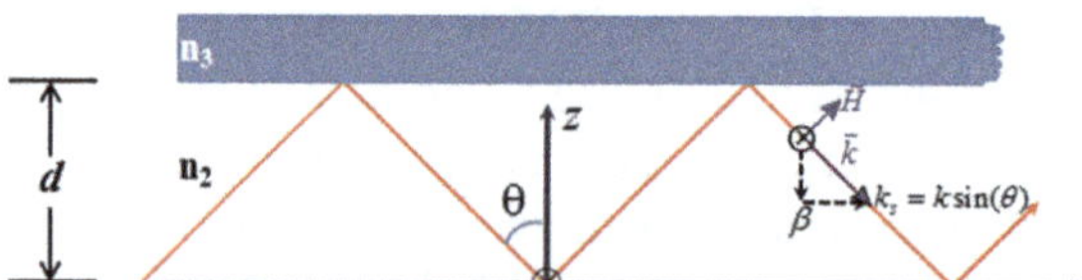

Figure 5.1. A slab dielectric waveguide constructed by a slab sandwiched by two cladding layers. A symbol ⊗ at the intersection of the x- and z-axes represents the y-axis pointing into the page. Another ⊗ at the intersection of the vector H and k indicates the electrical field of a TE wave parallel with the y-axis.

shown in (2.57), we have $E_y\big|_{z=0} = 0$ and $E_y\big|_{z=d} = 0$. From the first boundary condition at the interface $z = 0$, we get $A_2 = -A_1$, and from the second boundary condition at the interface $z = d$, we have $2jA_1 \sin[\beta d]e^{j[k_s x-\omega t]} = 0$. To satisfy the second boundary condition, we must have:

$$2\beta d = \frac{4\pi}{\lambda}\cos(\theta)d = 2m\pi \quad m = 1, 2... \tag{5.1}$$

This is the eigenvalue equation of the waveguide, where m is an integer called the 'mode index' with a limit of $m \leqslant 2d/\lambda$ set by $\cos(\theta) = 2m\pi/(\frac{4\pi}{\lambda}d) \leqslant 1$. The mode of the waveguide with a mode index m is the TE_m wave

$$E_y = 2jA_1 \sin[\beta z]e^{j[k_s x-\omega t]} = E_0 \sin\left(\frac{z}{d}m\pi\right)e^{j\left[2\pi\frac{x}{\lambda}\sin(\theta)-\omega t\right]} \tag{5.2}$$

where $E_0 = 2jA_1$ is also an unknown constant depending on the wave intensity. A waveguide only with a TE_1 mode is a single mode waveguide. The distance between the two metal plates of the single mode waveguide is: $\frac{\lambda}{2} \leqslant d < \lambda$. With the distance of the two plates slightly bigger than one wavelength, both the TE_1 wave and TE_2 wave can be transmitted in the system which is a multi-mode waveguide. Any TE_m wave with the mode index $m > 2$ that cannot be transmitted in the multi-mode waveguide is cut off. The incident angle of the electromagnetic wave propagating in a waveguide is discrete. For a single mode waveguide, the incident angle is $\theta = \arccos(\frac{\lambda}{2d})$. There is no such limit for TM waves propagating in the metal-walled waveguide. A former MIT professor Walter Lewin has a YouTube video [5] demonstrating a TE wave ($\lambda = 3$ cm) and a TM wave transmission through a metal-walled guide.

With knowledge about the metal-walled waveguide, we can handle a slab dielectric waveguide shown in figure 5.1. In the waveguide, there is a stationary wave in the slab. Evanescent waves exist in the top and bottom media. The electric fields of a TE wave in the system are assumed to be:

$$E(\vec{r}, t) = \begin{cases} E_{3y} \\ E_{2y} \\ E_{1y} \end{cases} = \begin{cases} E_{30}e^{-\gamma(z-d)}e^{j(k_s x-\omega t)} & (z > d) \\ E_{20}\cos(\beta z - \phi)e^{j(k_s x-\omega t)} & (0 < z \leqslant d) \\ E_{10}e^{\alpha z}e^{j(k_s x-\omega t)} & (z \leqslant 0) \end{cases} \tag{5.3}$$

where $\alpha = \sqrt{k_s^2 - n_1^2 k_0^2}$, $\beta = \sqrt{n_2^2 k_0^2 - k_s^2}$ and $\gamma = \sqrt{k_s^2 - n_3^2 k_0^2}$ are constants decided by the propagation constant k_s; E_{10}, E_{20} and E_{30} are unknown constants. From the boundary condition $\vec{n} \times (\vec{E}_2 - \vec{E}_1) = 0$ at the two interfaces of the slab, we have $(E_{2y} - E_{1y})\big|_{z=0} = 0$ and $(E_{2y} - E_{3y})\big|_{z=d} = 0$. These two conditions can be converted into the following two equations

$$E_{20}\cos(\phi) = E_{10} \tag{5.4}$$

$$E_{20}\cos(\beta d - \phi) = E_{30} \tag{5.5}$$

From Faraday's law: $\nabla \times \vec{E} = -\frac{\partial}{\partial t}\vec{B}$, it is easy to get the magnetic fields in the system in which the three media are non-magnetic materials ($\mu_r = 1$, $\vec{B} = \mu_0\vec{H}$)

$$\vec{H}(\vec{r}, t) = \begin{cases} \vec{H}_3 \\ \vec{H}_2 \\ \vec{H}_1 \end{cases} = \begin{cases} -\dfrac{j}{\mu_0\omega}[\gamma\vec{e}_x + jk_s\vec{e}_z]E_{3y} & (x > d) \\ -\dfrac{j}{\mu_0\omega}[\beta\tan(\beta z - \phi)\vec{e}_x + jk_s\vec{e}_z]E_{2y} & (0 < x \leqslant d) \\ -\dfrac{j}{\mu_0\omega}[-\alpha\vec{e}_x + jk_s\vec{e}_z]E_{1y} & (x \leqslant 0) \end{cases} \tag{5.6}$$

In the slab dielectric waveguide, there is no free current, so the boundary condition $\vec{n} \times (\vec{H}_2 - \vec{H}_1) = 0$ at the two interfaces of the slab can be written down exactly as $(H_{1x} - H_{2x})|_{z=0} = 0$ and $(H_{3x} - H_{2x})|_{z=d} = 0$. Using equations (5.4)–(5.6) at the interface $z = 0$, we get $\tan(\phi) = \frac{\alpha}{\beta} = \frac{\sqrt{n_2^2\sin^2(\theta) - n_1^2}}{n_2\cos(\theta)} = \tan(\frac{-\phi_{s21}}{2})$, where ϕ_{s21} is the phase angle shift of the TE wave totally reflected on the interface between medium 2 and medium 1 according to the formula shown in equation (4.14)

$$\phi = -\frac{1}{2}\phi_{s21} \tag{5.7}$$

Also at the interface $z = d$, we have $\tan(\beta d + \frac{1}{2}\phi_{21}) = \frac{\gamma}{\beta} = \frac{\sqrt{n_2^2\sin^2(\theta) - n_3^2}}{n_2\cos(\theta)} = \tan(\frac{-\phi_{s23}}{2})$, where ϕ_{s23} is the phase angle shift of the TE wave totally reflected on the interface between medium 2 and medium 3. Since the tangent function in trigonometry repeats every π, we can take the inverse tangent of both sides of the equation at the interface $z = d$, and multiply both sides by 2 to obtain an eigenvalue equation of the slab waveguide

$$\frac{4\pi d}{\lambda}n_2\cos(\theta) + \phi_{s21} + \phi_{s23} = 2m\pi \tag{5.8}$$

where m is an integer for the mode index of the slab dielectric waveguide. The eigenvalue equation of the TE wave transmitted in the waveguide is a transcendental equation. There are no explicit solutions for it. Discrete incident angles of TE waves transmitted in the waveguide can only be found numerically. The equation is the waveguide resonance condition of the slab waveguide for TE waves.

The resonance condition of the slab waveguide for TM waves can also be obtained in the same way. To get the eigenvalue equation of a TM wave transmitted in the slab dielectric waveguide, we can assume the magnetic fields in the system shown in figure 5.1 as:

$$H(\vec{r}, t) = \begin{cases} H_{3y} \\ H_{2y} \\ H_{1y} \end{cases} = \begin{cases} H_{30}e^{-\gamma(z-d)}e^{j(k_s x - \omega t)} & (x > d) \\ H_{20}\cos(\beta z - \phi)e^{j(k_s x - \omega t)} & (0 < x \leqslant d) \\ H_{10}e^{\alpha z}e^{j(k_s x - \omega t)} & (x \leqslant 0) \end{cases} \quad (5.9)$$

From $\nabla \times \vec{H} = \frac{\partial}{\partial t}\vec{D}$, we calculate the electric fields by $\vec{E} = \frac{1}{\varepsilon_r \varepsilon_0}\vec{D}$ in the system. Then from boundary conditions at the two interfaces, we have:

At $z = 0$, $H_{2y} = H_{1y}$, $H_{20}\cos(\phi) = H_{10}$

At $z = d$, $H_{3y} = H_{2y}$, $H_{20}\cos(\beta d - \phi) = H_{30}$

At $z = 0$, $E_{2x} = E_{1x}$, $-\frac{\beta}{n_2^2}H_{20}\sin(\phi) = -\frac{\alpha}{n_1^2}H_{20}\cos(\phi)$

At $z = d$, $E_{2x} = E_{3x}$, $\frac{\beta}{n_2^2}H_{20}\sin(\beta d - \phi) = \frac{\gamma}{n_3^2}H_{30} = \frac{\gamma}{n_3^2}H_{20}\cos(\beta d - \phi)$

It is easy to derive a transcendental equation from the above four boundary equations

$$\frac{4\pi d}{\lambda}n_2\cos(\theta) + \phi_{p21} + \phi_{p23} = 2m\pi \quad (5.10)$$

From formula (4.13), we know the phase angle shifts of total reflection of a TM wave at the two interfaces are $\phi_{p21} = -2\arctan\left(\frac{n_2^2}{n_1^2}\frac{\sqrt{n_2^2\sin^2(\theta) - n_1^2}}{n_2\cos(\theta)}\right)$ and $\phi_{p23} = -2\arctan\left(\frac{n_2^2}{n_3^2}\frac{\sqrt{n_2^2\sin^2(\theta) - n_3^2}}{n_2\cos(\theta)}\right)$ respectively.

5.2 The benchmark of a RM biosensor: IAsys

IAsys is a waveguide biosensor. As shown at the lower-right corner of figure 5.2, directly deposited on the base of a glass prism ($n_0 = 1.8$) is a coupling layer made of silica (SiO_2, $n_1 = 1.46$, $d_1 = 500$ nm), and over the silica film is a resonant layer made of titania (TiO_2, $n_2 = 2.2$, $d_2 = 100$ nm). On the interface between the titania and water (n_4), biological molecules assemble into a film with unknown parameters (n_3, d_3). The principle section of the polarizer is set at 45° (φ) and the principle section of the analyzer was set at −45° (ψ). A linear polarized light from the polarizer is reflected on the prism/silica interface with an incident angle θ bigger than the critical angle ($\theta_c = 54.205°$) of the system.

An evanescent wave could be coupled into the titania film which works as a waveguide [4]. In the titania film, light propagates in a zigzag pathway with an incident angle $\theta_2 = \arcsin[\frac{n_0}{n_2}\sin(\theta)]$. Total reflections also occur on the interface between the titania film and the biological film. An evanescent wave penetrates into the sample layer. However, only at some incident angles satisfying resonant conditions, does efficient coupling occur. The resonance angle of the system is sensitive to the sample film.

There are two s–p Cartesian coordinate systems in figure 5.2. Looking back into the light, the left s–p plane is perpendicular to the incident light, and the right s–p plane is

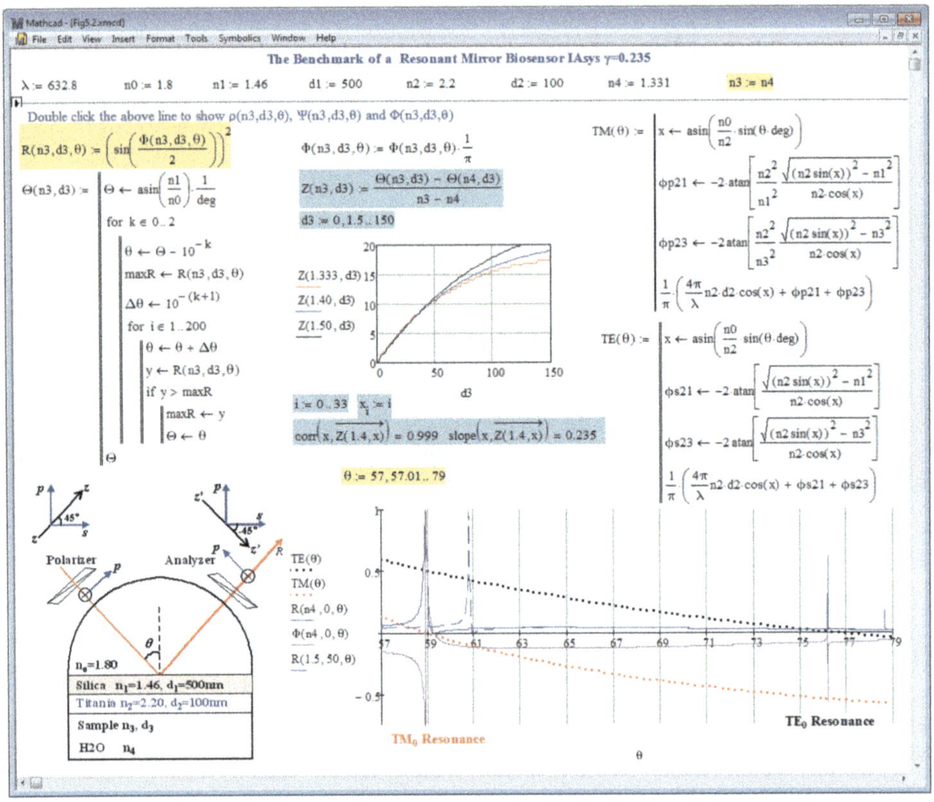

Figure 5.2. Mathcad program for numeric simulations of an IAsys biosensor. The benchmark of the system is 0.235, and it is valid for the sample film with a thickness up to 33 nm. The symbol ⊗ on the light rays represents two *s*-axes. The program is available online at http://iopscience.iop.org/book/978-0-7503-2384-0.

perpendicular to the reflected light. The two *p*-axes are perpendicular to the light rays respectively, so the two *s*-axes are parallel with each other and point into the page. The angle of the principle section of the polarizer to the left *s*-axis is $\varphi = 45°$, and the angle of the principle section of the analyzer to the right *s*-axis is set to be $\psi = -45°$. Passing through the polarizer, the incident light becomes a linear polarized light with A as its electric field amplitude. Before being reflected at the reflection point on the base of the prism, the polarized light's two components in the left *p*-axis direction and in the left *s*-axis direction have the same amplitude and the same phase

$$\begin{cases} E_{pi} = A\sin(\varphi)e^{j(kr-\omega t)} = \dfrac{A}{\sqrt{2}}e^{j(kr-\omega t)} \\ E_{si} = A\cos(\varphi)e^{j(kr-\omega t)} = \dfrac{A}{\sqrt{2}}e^{j(kr-\omega t)} \end{cases} \tag{5.11}$$

Then the TM wave and the TE wave of the incident light are reflected by the multilayer system. On the left face of the analyzer, the two components of the reflected light are:

$$\begin{cases} E_{pr} = \dfrac{A}{\sqrt{2}} r_p e^{j(kr-\omega t)} \\ E_{sr} = \dfrac{A}{\sqrt{2}} r_s e^{j(kr-\omega t)} \end{cases} \tag{5.12}$$

Both the TM wave and the TE wave can be decomposed into two components in front of the analyzer. Only two components ($E_{pr}\cos(\frac{\pi}{4})$ and $E_{sr}\cos(\frac{\pi}{4})$) parallel with the principle section of the analyzer ($z'z'$) can pass through the analyzer, but in opposite directions. Using an arrow as the positive direction of the $z'z'$-axis on the right s–p Cartesian system, we obtain the electric field of the reflected light exiting from the analyzer

$$E_r = \frac{A}{2}(r_s - r_p)e^{j(kr-\omega t)} \tag{5.13}$$

In total reflection, we have $|r_p| = 1$ and $|r_s| = 1$ (see equations (4.11)–(4.14)). According to the formula (2.28), the reflected light intensity (I_{-45}) received by the detector of the system is proportional to $|E_r|^2$. When the incident light penetrated in and out of the prism and the analyzer, a small fraction of the energy would be lost. If the light rays are perpendicular to the surfaces of the two optical elements, there is no difference in the energy losses between the TM and TE waves. A parameter η ($\leqslant$ 100%) in the following formula is used for the main fraction of light energy received by the detector

$$I_{-45} = \frac{1}{4}\left| r_s - r_p \right|^2 \eta I_0 = \sin^2\left(\frac{\Phi}{2}\right)\eta I_0 = R(\theta)I_0 \tag{5.14}$$

In the above formula I_0 is proportional to A^2, $R(\theta) = \eta\sin^2(\frac{\Phi}{2})$ is the reflectivity of the RM system, and $\Phi = \phi_p - \phi_s$ is the traditional ellipsometry parameter: the phase angle (the phase angle difference between the TM wave and the TE wave). A RM spectrum is a curve about the relationship between the light reflectivity $R(\theta)$ and the incident angle θ. An IAsys biosensor uses an angle scan system to measure RM spectra.

A Mathcad program for numeric simulations of the IAsys biosensor is presented in figure 5.2. In the program, η was set to be 100%. Any η value ($\leqslant 1$) would not affect the simulation results since we only searched resonant angles (peak positions) of RM spectra in the program. All parameters listed in the second line of the program came from Cush's paper [1]. Below these parameters three sub-programs $\rho(n3, d3, \theta)$, $\Psi(n3, d_3, \theta)$ and $\Phi(n3, d3, \theta)$ collapsed into a line with an arrow on its left. The phase angle function $\Phi(n3, d3, \theta)$ is the same function defined in the Mathcad program shown in figure 4.7, but here the unit of Φ is in radians, not in degrees. Also the first line of that function is changed to $n = (n0, n1, n2, n3, n4)^T$ to match with those variable names in this program. The text region under the long line instructs readers how to view those hidden functions on the Mathcad programming interface. Then the reflectivity function of the system $R(n3, d3, \theta)$ is defined according to the formula (5.14). The value of the function is proportional to the intensity of the reflected light exiting from the analyzer. The unit of the sub-programs $\Phi(n3, d3, \theta)$ is in radians, and it is then normalized to π so that it can be plotted with $R(n3, d3, \theta)$ in a same x–y plot with the same scale.

On the right of the program interface, function $TM(\theta)$ and function $TE(\theta)$ are defined according to the left sides of equations (5.10) and (5.8) respectively. Both are normalized by π. Two curves with dots on the x–y plot at the lower-right corner of the program are outputs of the two functions. At $\theta = 58.917°$, we see the red dot line crosses the x-axis: $TM(\theta) = 0$, so there is an optical resonance in the waveguide system for the TM_0 wave. Also at $\theta = 76.168°$, there is another optical resonance in the waveguide system for the TE_0 wave. A thin blue solid curve is the RM spectrum for $R(n4, 0, \theta)$ with its peaks at 58.917° and 76.168° respectively. The purple solid curve is the output of $\Phi(n4, 0, \theta)$. The ellipsometry phase angle Φ flips over at both the TM_0 and TE_0 resonance angles. The dashed blue curve is an RM spectrum $R(1.5, 50, \theta)$. It shows that a thin film ($d3 = 50$ nm, $n3 = 1.5$) attached to the titania film in water shifts the TM_0 wave and TE_0 wave resonance angles to 60.785° and 78.656° respectively. The TE_0 resonance peaks are so sharp that they are almost vertical straight lines. It is convenient to focus on the TM_0 resonance peaks and we use the symbol Θ to represent the resonance angle position: the incident angle corresponding to a TM_0 peak on a RM spectrum.

On the left of the program interface, a sub-program $\Theta(n3, d3)$ is defined above the sketch of the RM optical system. It is used to investigate how the parameters of the sample film affect the TM_0 resonant angle position. This sub-program is almost the same as the one shown in figure 3.5 for searching for the θ_{SPR} of the SPR spectra. Since we are searching the positions of RM spectra peaks, the 'if' condition in the sub-program is changed from '$y \leqslant$ mini$R$' to '$y >$ max$R$'. One more sub-program $Z(n3, d3)$ is defined. Three curves in an x–y plot in the middle of the program interface are outputs of $Z(1.333, d3)$, $Z(1.4, d3)$ and $Z(1.5, d3)$ respectively while $d3$ is changing from zero to 150 nm. The benchmark of the RM system is found to be $\gamma = 0.235$ [deg nm^{-1}], and the linear region of the linear relationship is up to $d3 = 33$ nm with a correlation coefficient bigger than 99.9%. According to the formula (1.12), a linear relationship between the shift of the resonant angle ($\Delta\Theta$) and the change of the sample surface concentration ($\Delta\Gamma$) is obtained. The linear response region of the RM system is $d3_{max} = 33$ nm, or $\Delta\Theta_{max} = 1.311°$, or $\Delta\Gamma_{max}$=31.2 ng mm^{-2}

$$\Delta\Theta = 0.042\Delta\Gamma \tag{5.15}$$

Corresponding to $\Delta\Gamma = 1$ ng mm^{-2}, BIAcore 1000, a commercial SPR biosensor's resonance angle shift is 0.1°, but the RM biosensor's is only 0.042°. The sensitivity of the RM biosensor is about 42% of that of BIAcore 1000. Since the angle unit degree is too big for RM biosensors, arc second (1°/3600) is used as a unit for their resonant angle shift. Rhodes [6] used an RM biosensor to study interactions between Tamm–Horsfall protein (THP) and complement 1q (Cq1). THP is one of the most abundant proteins in normal human urine. Its molecular weight is 86 kDa, and Cq1 (480 kDa) is a protein of the innate immune system. The maximum RM resonant angle shift was less than 200 arc sec (0.056°) for the THP. Thermo BioAnalysis (Santa Fe, NM), the company making RM biosensors had to use a dextran matrix over the titania film to improve the sensitivity of RM biosensors. But using a dextran matrix to immobilize biomolecules on sensing chips infringed a patent of Biacore (Uppsala, Sweden). After Thermo BioAnalysis lost a lawsuit [7] in December 30, 1999, RM biosensors disappeared from the market.

5.3 RMEP and RMEPA, two super biosensors

Monitoring reflected light intensities at a fixed incident angle while biological molecules are assembling on the titania film of a sensing chip, we can extract Φ information with formula (5.14). In this way, optical biosensors with a monochromatic light source would have a very simple structure, because even a mechanical system for an angle scan would not be needed. Since the new biosensor can measure ellipsometry parameters Φ and Ψ (= 45°), it is named 'RMEP', an abbreviation of 'resonant mirror' and 'ellipsometry'. To use that formula, we have to find the maximum reflected light intensity ($I_{\max} = \eta I_0$) received by the detector

$$\Phi = 2\text{arc}\sin\left(\sqrt{\frac{I_{-45}}{I_{\max}}}\right) \tag{5.16}$$

There is a simpler way to find the value of $I_{\max}$. If the analyzer's principle section could be rotated 90° to $\psi = 45°$, the reflected light intensity of the RM system could be derived from the following formula, where I_{+45} is used to distinguish it from the RM system's reflected light intensity I_{-45}

$$I_{+45} = \frac{1}{4}\mid r_s + r_p \mid^2 \eta I_0 = \cos^2\left(\frac{\Phi}{2}\right)\eta I_0 \tag{5.17}$$

At any incident angle θ of light in total reflection, we only need to measure reflected light intensities $I_{-45}(\theta)$ and $I_{+45}(\theta)$ by rotating the analyzer in two positions and the sum of them is a constant

$$I_{-45} + I_{+45} = \eta I_0 = I_{\max} \tag{5.18}$$

If the light intensity of a monochromatic light source in a RMEP biosensor is stable enough, only at the beginning and at the end of an affinity assay experiment, we need to measure both I_{-45} and I_{+45} for an average light intensity $I_{\max}$ to normalize a I_{-45} spectrum at a fixed angle Θ

$$I_{\max} = \frac{1}{2}\left[(I_{-45} + I_{+45})_{begin} + (I_{-45} + I_{+45})_{end}\right]\Bigg|_{\theta=\Theta} \tag{5.19}$$

With the $I_{\max}$ to normalize I_{-45} in equation (5.16), we could miss information about negative phase angle values ($-180° \leqslant \Phi < 0°$). A condition has to be added to equation (5.16) to extract the correct phase angle. From the RM spectra shown in figure 5.2, we need to set $\Phi = -2\text{arc}\sin\left(\sqrt{\frac{I_{-45}}{I_{\max}}}\right)$ if $\frac{d}{d\theta}(I_{-45}) \geqslant 0$.

A Mathcad program for numeric simulations of a virtual RMEP biosensor is shown in figure 5.3. In the RMEP biosensor, we need 'broad' RM spectra, so the coupling film thickness is changed from 500 nm to 120 nm, and the resonant film thickness is reduced from 100 nm to 80 nm. All parameters of the new biosensor are listed in the second line of the program. With these parameters, a sub-program $\rho(n3, d3, \theta)$ is defined for ellipsometry. It is a ratio of $rp_{0,4}$ and $rs_{0,4}$, Fresnel coefficients of TM wave and TE wave respectively. Ellipsometry parameter functions $\Phi(n3, d3, \theta)$ and $\Psi(n3, d3, \theta)$ are then derived from $\rho(n3, d3, \theta)$ as usual. These functions are hidden in a line and they can be shown by double clicking the line on the Mathcad program interface.

Under the long line, an RM reflectivity function $R(n3, d3, \theta)$ is defined for I_{-45} according to the formula (5.14), and $M(n3, d3, \theta)$ is defined for I_{+45} according to the

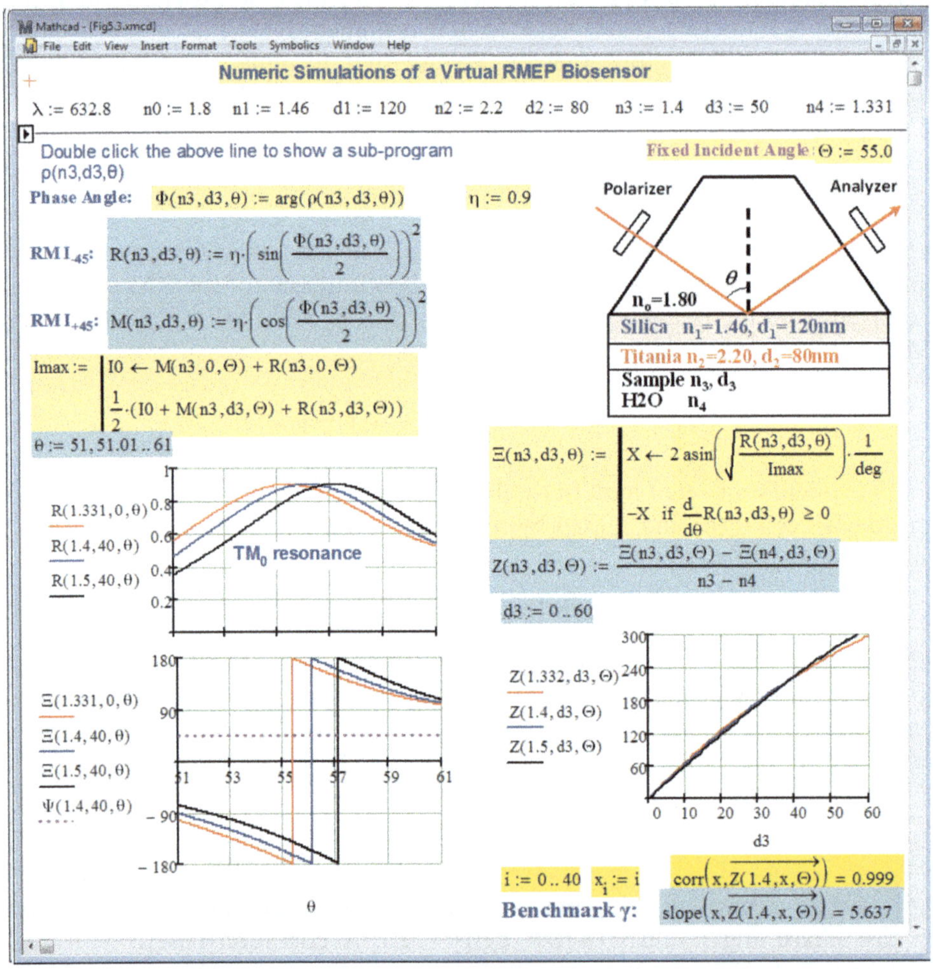

Figure 5.3. A numeric simulation program for a possible super optical biosensor RMEP. The benchmark γ of the super biosensor is increased to 5.637 deg nm^{-1} with a linear region up to $d3_{max} = 40$ nm, or $\Delta\Phi_{max} = 38.1°$, or $\Delta\Gamma_{max} = 37.5$ ng mm^{-2}. The program is available online at http://iopscience.iop.org/book/978-0-7503-2384-0.

formula (5.17). In the two formulae, η was set to be 90% (could be another constant). Three RM spectra $R(1.331, 0, \theta)$, $R(1.4, 40, \theta)$ and $R(1.5, 40, \theta)$ are shown in an x–y plot on the left side of the program. These RM spectra are the results of TM_0 resonance. In the program, the fixed incident angle Θ was set to be 55.0°. With the I_{max} to normalize I_{-45}, we extracted $\Phi(n3, d3, \theta)$ in the program according to the formula (5.16) with an extra condition to set Φ to be a negative number if $\frac{d}{d\theta}(I_{-45}) \geqslant 0$. The extracted phase angle was redefined as $\Xi(n3, d3, \theta)$ under the schematic of the RM system in the figure. Φ and Ξ should be the same. Three $\Xi(n3, d3, \theta)$ spectra are shown in an x–y plot at the lower-left corner of the program interface. An amplitude angle curve is also shown in the plot. It is the horizontal dotted line with $\Psi = 45°$.

For the incident angle in the region of $54° \leqslant \theta \leqslant 55.6°$, the Ξ or Φ spectra are almost parallel to each other. This is an advantage of RMEP over SPREP, because the benchmark of a RMEP system is not sensitive to the incident angle. For a given θ (= Θ) in the region, one more function $Z(n3, d3, \Theta)$ was defined to investigate how $\Delta\Xi$ responds to the change of the sample's effective optical thickness. The benchmark (γ = 5.637 deg nm^{-1}) and the linear region ($d4_{max}$ = 40 nm) of the virtual RMEP come from the straight lines in the x–y plot on the lower-right corner of the program. From those data, we could see the linear region of the RMEP was up to $\Delta\Phi_{max}$ = 38.1°, or $\Delta\Gamma_{max}$ = 37.5 ng mm^{-2}. The virtual RMEP biosensor with parameters shown in the program is about 10 times as sensitive as BIAcore 1000

$$\Delta\Phi = 1.01\Delta\Gamma \qquad (5.20)$$

Formula (5.16) provides a simple and fast method to measure Φ for the RMEP biosensor. By just monitoring reflected light intensities at a fixed incident angle Θ, we could extract Φ information, and $\Delta\Phi$ is proportional to $\Delta\Gamma$. Since no angle scan is needed, the structure of the RMEP biosensor is simpler than any SPR, EP, or RM biosensor.

In a possible structure of the RMEP biosensor, the prism with a sensing chip should be held on a rotatable stage with an electric angle sensor to measure the incident angle. By rotating the stage we can easily find the sign of $\frac{d}{d\theta}(I_{-45})$ for the phase angle extractions. The light source of an RMEP biosensor should be very stable, and its detector should linearly respond to light intensity. A light intensity stabilized He–Ne laser (Melles Griot model: 25 STP 912) could be a choice. It is a 1 mW compact laser system with a coherent length in kilometers. According to the product information provided by the company, over 8 h the laser's output power fluctuation is less than 0.2%, and its polarization extinction ratio is 5000 : 1. For the light detector, a photomultiplier tube or a solar cell should be replaced with a photodiode whose output is proportional to light intensity. A Si photodiode made by Hamamatsu (model: S1227-1010BR) could be a choice. The detector has an active area of 10 × 10 mm^2 with a spectral response range from 340 to 1000 nm. Its peak sensitivity wavelength is at 700 nm. With low dark current, the photodiode is suitable for our optical biosensor: RMEP.

A disadvantage of the RMEP biosensor is that the prism with a coupling film and a resonant film has to be discarded every time after a test is finished. Such a glass prism sensing system is more expensive than a coverslip with a gold film in SPR biosensors. It is possible to make the prism systems affordable by using an optical plastic MR-174 made by Japan's Mitsui Chemicals for eye glasses. Its refractive index is 1.74. With a plastic prism, and at a fixed incident angle of $\theta = 58.5°$ in numeric simulations, we get the benchmark of the new biosensor: $\gamma = 5.634$ deg nm^{-1}.

We can even build a more powerful optical biosensor by a hybridization of RMEP with attenuated total reflectance (ATR), as shown in figure 5.4. The new biosensor is named 'RMEPA'. In a schematic of the RMEPA system shown in the figure, there is a plastic trapezoid prism with two equal short sides where light gets in and out perpendicularly. Each of its two sharper angles is 58.5°. On the top and

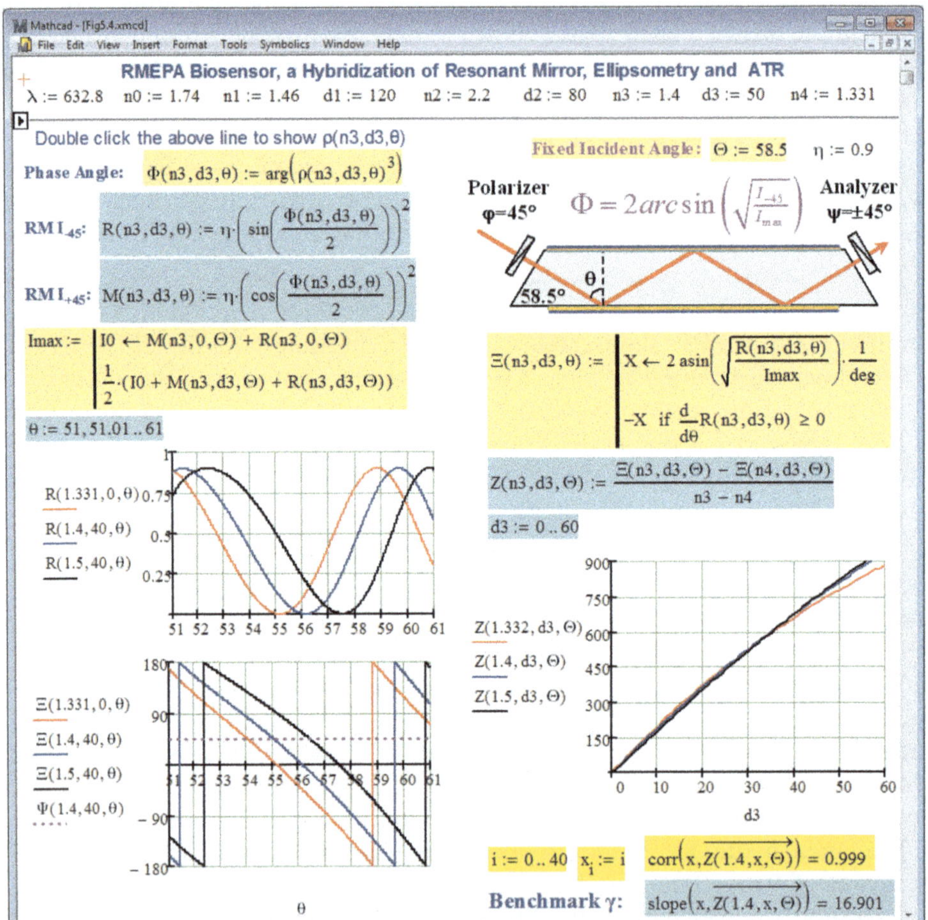

Figure 5.4. Mathcad program to simulate a virtual super biosensor RMEPA, a hybridization of resonant mirror, ellipsometry and ATR. Its benchmark is 16.9, and its linear responding region is up to $\Delta\Phi_{max} = 114.3°$, or $\Delta\Gamma_{max} = 37.6$ ng mm^{-2}. The program is available online at http://iopscience.iop.org/book/978-0-7503-2384-0.

bottom of the trapezoid prism, there is a silica film and a titania film. Their thicknesses are 120 and 80 nm respectively. After the linear polarized He–Ne laser perpendicularly gets into the left surface of the prism, it is reflected three times inside the prism with an incident angle of 58.5°. From numerical simulations, we get the RMEPA's benchmark $\gamma = 16.9$ deg nm^{-1}, and the linear relationship about the phase angle change ($\Delta\Phi$) and the sample surface concentration increment ($\Delta\Gamma$).

$$\Delta\Phi = 3.04\Delta\Gamma \tag{5.21}$$

The linear responding region of the RMEPA biosensor is found to be up to 114.3°, or $\Delta\Gamma_{max} = 37.6$ ng mm^{-2}. The minimum length of the trapezoid prism would be 24.5 mm if the prism thickness was 5 mm. By increasing the prism length for five total reflections of the light inside the prism, the benchmark of the RMEPA system could be $\gamma = 28.2$ deg nm^{-1} with the same linear region (40 nm). If optical fibers could be used to replace the prism, the RMEPA biosensor could be simpler.

Without modification, a traditional null ellipsometer could be used to make a super biosensor RMEP or REMPA. The fast axis or e-axis of the $\lambda/4$ plate should be set at $\vartheta = 45°$, and the analyzer at $\psi = -45°$ (figure 4.2). Placing a prism with a coupling film and a resonant film on to a sample stage of the null ellipsometer, we only need to rotate the polarizer's angle (φ) at a fixed incident angle θ to extinguish the reflected light received by the photomultiplier tube. Practically, at the beginning of an affinity assay, we could manually rotate the polarizer to extinguish the reflected light received by the light detector. The polarizer's angle was recorded as φ_0. After ligand–receptor interactions start, the polarizer's angle φ would change with time for a null reflected light intensity, and we have $\Delta\Phi = 2|\varphi - \varphi_0|$ according to equation (4.4). Since we only need to adjust the polarizer in a traditional null ellipsometer, manual measurements could be fast enough. We could measure $\Delta\Phi$ automatically by using a step motor to rotate the polarizer connected to an electric angular sensor.

5.4 A porous silicon interferometric biosensor

In human history, there was the Stone Age, the Bronze Age and the Iron Age. Now we are in the Silicon Age. Since the 1950s, electronic devices such as computers and cell phones, based on silicon IC chips, have greatly changed our lives. In 1997, there was a paper about the most sensitive biosensor using porous silicon (Psi) [8]. The name of the biosensor was the 'PSi interferometric sensor' or the 'PSi Biosensor'. The paper claimed that the detection limit of the sensor for DNA concentration was 9×10^{-6} ng mm^{-2}, about 300 times as sensitive as SPR biosensors. Interferometric biosensors are not waveguide biosensors. Only when the incident angle of light in the waveguide shown in figure 5.1 is zero, will light waves bound up and down in the slab to produce interference patterns. There is no light energy propagating in the slab.

A photo [9] of a cross section of a porous silicon film taken by a scanning electron microscope is shown in figure 5.5. The thickness of the PSi film was 3004 nm. The inset in the upper right corner of the photo is a photo of the porous silicon film

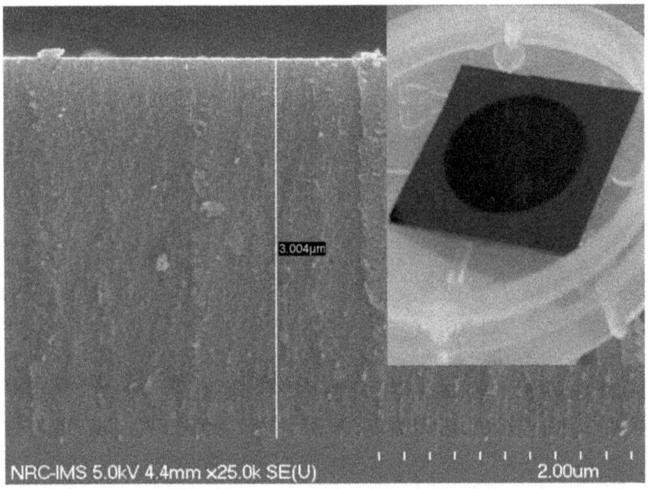

Figure 5.5. A Scanning electron micrograph of a porous silicon sample (cross-sectional view). The inset is a photo of the porous silicon film etched on a piece of silicon wafer (20 × 20 × 2 mm^3) in a white plastic container. Reprinted with permission from [9]. Copyright (2002) American Chemical Society.

etched on a piece of silicon wafer (20 × 20 × 2 mm^3) in a white plastic container. Heavily doped p-type silicon wafers (⟨100⟩, boron-doped, 0.0018–0.0022 Ω·cm, International Wafer Service) were used to prepare porous silicon samples by an anodic etch in an ethanolic HF solution (HF/Ethanol 3:1 v/v) made from 48% HF in water. Prior to the etching procedure, the wafers were cleaned with a 3:1 (v/v) of H_2SO_4/H_2O_2 for 30 min, rinsed with copious amounts of Milli-Q water, and then immersed in the HF solution for 60 s. The wafers were dried under nitrogen gas, contacted on the backside with an In/Ga eutectic, and mounted in a Teflon etching cell. Porous silicon layers were fabricated using a two-electrode setup at a constant current of 250 mA cm^{-2} for 10 s in the absence of light using a Pt mesh counter electrode to ensure a homogeneous electric field. The etched wafer was then removed from the cell, rinsed with ethanol, and dried under nitrogen.

The freshly prepared PSi surface was placed under argon in a Schlenk tube containing a deoxygenated neat 1-decene or ethyl-undecylenate and allowed to react at 115 °C for 16 h. The excess of unreacted reagent was removed by rinsing at room temperature, with tetrahydrofuran and 1,1,1-trichloroethane, and then the sample was dried under a stream of nitrogen. SEM images were obtained with a Hitachi S4700 field emission microscope (Hitachi, Tokyo, Japan) using an accelerating voltage of 5.0 keV. Top and cross-sectional views were obtained at varying magnifications to examine the pore morphology.

Suppose the thickness of the PSi film is ℓ in [nm], and biological receptor molecules were homogeneously immobilized in the PSi film with a concentration C [g cm^{-3}], the surface concentration of the biological sample should be $\Gamma = \ell \cdot C$ in units of [ng mm^{-2}]. Ligand molecules in water specifically binding with receptor molecules in the PSi film would increase the biological samples concentration:

$\Delta\Gamma = \Delta(\ell \cdot C) = \frac{\partial C}{\partial n}\Delta(\ell \cdot n)$, where ℓ is a constant, and n is the refractive index of the PSi film with biological samples at a wavelength λ in water. When white light from optical fibers was vertical shining on the PSi film in water, light rays reflected from the water/PSi interface and the PSi/Si interface would interfere with each other to create a Fabry–Perot interference pattern which was determined by the effective optical thickness $(\ell \cdot n)$ of the film. We know that a constructive interference creates a peak, and a destructive interference creates a valley in an interference pattern. A peak at a wavelength has an interference order m (an integer). According to the formula (3.9), the phase difference of reflected light rays from the two interfaces of the PSi film is $\frac{2\pi}{\lambda}(2n\ell)$. For a peak at λ, the phase difference is equal to $2m\pi$. A small increase of n would not change the interference order, but would shift the pattern to a longer wavelength. From $\frac{2\pi}{\lambda}(2n\ell) = 2m\pi$, we get the benchmark of the PSi biosensor: $\gamma = \frac{2}{m}$, a dimensionless constant

$$\Delta\lambda = \frac{2}{m}\Delta(n \cdot \ell) = \frac{2}{m}\frac{\partial n}{\partial C}\Delta\Gamma \tag{5.22}$$

If the interference order m is equal to 20, then we have the benchmark $\gamma = 0.1$, and $\Delta\lambda = 0.018\Delta\Gamma$. In the case when the λ resolution is 0.01 nm for the PSi optical system, the low limit of the PSi biosensor for biological samples should be 0.56 ng mm^{-2}. It is not as sensitive as SPR biosensors. Freshly etched porous silicon is unstable in water because of oxidation and hydrolytic corrosion. Another paper [10] published by another group in the same laboratory [8] pointed out these problems. The authors of that paper estimated that the low limit of protein detections would to be 1–10 ng mm^{-2}.

It is easy to find the orders of an interference pattern if the wavelengths of the interference pattern are extended to 2000 nm. The refractive indices of silicon at wavelengths beyond 1250 nm could be treated as real numbers around 3.5. Figure 5.6 shows a real interference pattern [9] of the PSi film in the air. Interferometric reflectance spectra of a porous silicon layer were obtained using an IR spectrometer (Control Development, South Bend, IN) in which an optical fiber bundle is trifurcated and serves to direct incandescent light to the sample as well as leading the reflected light back to the detectors (a Si linear array and an InGaAs array, made by Control Development). The use of these detectors allows two stages of normalization before the reflectance spectrum was taken from the porous silicon sample. The first is subtraction of a dark spectrum (signal obtained with no light present), and the second is normalization to a reference spectrum (Spectralon, Labsphere). The illumination and detection of the reflected light was performed using an incident angle normal to the PSi surface in the range of 380–2200 nm.

Suppose the order of the peak on the far right of the figure is m with the wavelength λ_2, the orders of two neighbor valleys on the left and right of the peak should be $(m + \frac{1}{2})$ and $(m - \frac{1}{2})$ respectively, so we have three equations: $2(m + \frac{1}{2})\pi = \frac{4\pi n\ell}{\lambda_3}$, $2m\pi = \frac{4\pi n\ell}{\lambda_2}$, and $2(m - \frac{1}{2})\pi = \frac{4\pi n\ell}{\lambda_1}$ where λ_1 and λ_3 are the

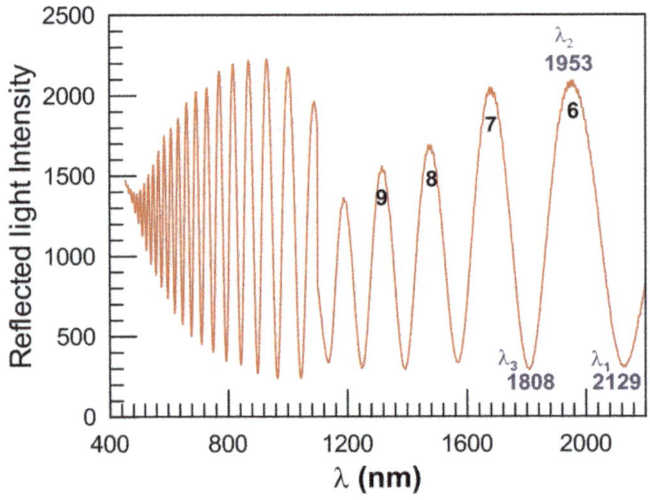

Figure 5.6. A Fabry–Perot interference pattern of a PSi film in the air (the interference order of last peak on the right is found to be 6 by calculations with the three wavelengths shown in the figure). Reprinted with permission from [9]. Copyright (2002) American Chemical Society.

wavelengths corresponding to the two valleys respectively. From the above three equations, we obtain a formula to calculate the interference order of the last peak on the right of the interference pattern. With the three wavelengths shown in figure 5.6, the order m was found to be 6.14. Since m should be an integer, it is reasonable to be set as $m = 6$. Then it is easy to get all order numbers of those peaks from right to left. At $\lambda = 600$ nm, a peak order is 20

$$m = \frac{1}{\lambda_2}\frac{\lambda_1\lambda_3}{\lambda_1 - \lambda_3} \tag{5.23}$$

References

[1] Cush R, Cronin J M and Stewatr W J 1993 The resonant mirror: a novel optical biosensor of direct sensing of biomolecular interactions. Part I: Principle of operation and associated instrumentation *Biosens. Bioelectron.* **8** 347–53

[2] Hodgson J 1994 Light, angles, action instruments for label-free, real-time monitoring of intermolecular interactions *Biotechnology* **12** 31–5

[3] Buckle P E, Davies R J, Kinning T, Yeung D, Edwards P R and Pollard-Knight D 1993 The resonant mirror: a novel optical sensor for direct sensing of biomolecular interactions Part II: Applications *Biosens. Bioelectron.* **8** 355–63

[4] Tien P K 1971 Light waves in thin films and integrated optics *Appl. Opt.* **10** 2395–413

[5] https://www.youtube.com/watch?v=Fmwu6AKPpBI&t=2512s

[6] Rhodes D C J 2000 Binding of Tamm-Horsfall protein to complement 1q measured by ELISA and resonant mirror biosensor techniques under various ionic-strength conditions *Immunol. Cell Biol.* **78** 474–82

[7] https://ipmall.law.unh.edu/sites/default/files/hosted_resources/Markman/pdfFiles/1999.12.30_BIACORE_AB_v._THERMO_BIOANALYSIS_CO.pdf
[8] Lin V S-Y, Motesharei K, Dancil K-P S, Sailor M J and Ghadiri M R 1997 A porous silicon-based optical interferometric biosensor *Science* **278** 840–3
[9] Xiao C, Boukherroub R, Wojtyk J T C, Wayner D D M and Luong J H T 2002 Structure characterization of porous silicon layers based on a theoretical analysis *Langmuir* **18** 4165–70
[10] Janshoff A and Dancil K-P S *et al* 1998 Macroporous p-type silicon fabry–perot layers. fabrication, characterization, and applications in biosensing *J. Am. Chem. Soc.* **120** 12108–16

Chapter 6

SPR biosensors DIY and applications

Information about three homemade SPR biosensors and their applications is presented in this last chapter. The first one is based on a spectrometer; the second one is based on a goniometer of an x-ray diffraction; and the third one is based on a capacitance angle sensor. With this information, readers will know how to choose a light source, how to choose glass to make prisms, how to deposit gold film on a glass chip, how to trace reflected light for incident angle scan of SPR spectra, and how to write programs for data acquisitions, etc.

Formulae about light reflections and refractions in multilayer optical systems are well known. With those formulae, we can write Mathcad programs to make virtual biosensors based on the principles of surface plasmon resonance, ellipsometry, resonant mirror, and attenuated total reflection. Hybridizations of these four optical techniques made it possible to build super biosensors: SPREP, RMEP and RMEPA. In cyberspaces, we investigated benchmarks of biosensors without knowing how to precisely measure incident angles, record light intensities, deposit gold films on glass chips, etc. It is possible to make a real optical biosensor ourselves, especially a surface plasmon resonance biosensor with Kretschman configuration. This kind of optical instrument is simple and robust. After Liedberg and his colleagues [1] published a paper about biosensing applications of the surface plasmon resonance technique, SPR biosensors became popular. Three homemade SPR biosensors are presented in this chapter.

6.1 Modifying a spectrometer to make a simple SPR biosensor

To make an SPR biosensor, we need a glass prism, a monochromatic light source (laser), a polarizer, a light detector, a signal recorder, and a goniometer. All of these should be integrated into an instrument. A second-hand prism spectrometer shown in figure 6.1 was used as the platform for all the above components. The spectrometer had a rotatable black table to hold the prism, a collimator on a fixed

doi:10.1088/2053-2563/ab0823ch6

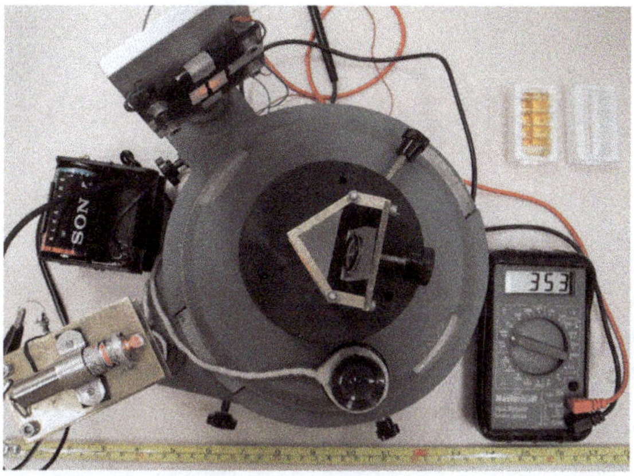

Figure 6.1. A simple SPR biosensor built on a spectrometer with an angle resolution of 1′ (= 1°/60). The light source is a semiconductor laser pointer emitting a linear polarized laser ($\lambda = 650$ nm). A solar cell was used to detect light signals displayed by a multimeter.

arm, and a telescope on a rotatable arm. But the collimator was removed so a semiconductor laser pointer could be fixed there as a light source. Also, the telescope was removed, and a solar cell of size 10×10 mm^2 was set there as a light detector. Under the prism table there was a big horizontal disk fixed on the heavy base of the spectrometer. On the rim of the disk there was a scale to divide the rim into 720 segments. The angular resolution of the rim scale was 0.5° or 30 min. There were four Vernier scales on the rim, and each of them had an angular resolution of 1 min. Two of them on two ends of the diameter of the rim were for the prism table, and the other two were for the light detector on the rotatable arm. In the photo, a small optical lens amplifier attached to the end of a thread was used to read the Vernier scales. By rotating the prism table, the incident angle of the SPR system could be changed. A lock-screw could lock the incident angle, so the reflected light from the prism could be traced by the detector on the rotatable arm which could also be locked in any position.

The output power of the laser was 1 mW and the laser wavelength was 650 nm. Since the laser is a linear polarized light, a polarizer is not needed for the simple SPR system. Its light spot was a strip with the ratio of its height to width of about 10. The strip was parallel with the p–n junction plane of the laser diode. The linear polarization plane of the laser was perpendicular to the strip of the laser spot. If all the components of the SPR system were set on a horizontal plane as shown in the photo, we could adjust the red laser to be a p-polarized light if the laser strip was a vertical line. The laser diode was powered by batteries (1.5 V × 3). A lens in front of the laser diode collimates the laser beam into a round spot. The output of the solar cell was 500 mV if the laser was directly shining on it. The output of the solar cell measured by a multimeter was 353 mV when the photo in the figure was taken.

The prism was made of ZF3 glass. A top view of the prism was a 90°–45°–45° triangle. The sizes of the prism were L25 × W25 × H25 in mm. The prism was held on the sample table of the spectrometer by a brass frame shown in figure 6.2(A).

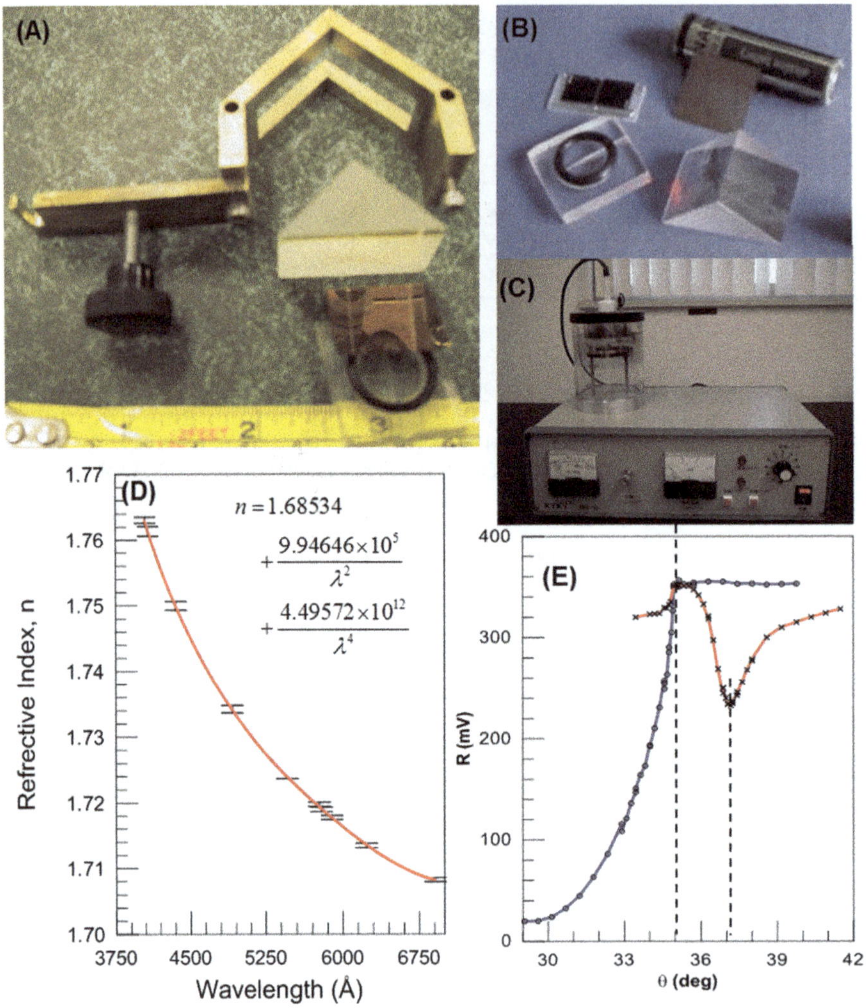

Figure 6.2. (A) A brass frame to hold the prism on the table of the spectrometer. (B) A type AAA battery was used as a size reference for the prism, the gold film deposited on a coverslip (20 × 20 × 0.2 mm^3), two solar cells and a rubber O-ring glued on a piece of Plexiglas. (C) A vacuum ion sputter was used to deposit gold films on glass. (D) A dispersion curve of ZF3 glass. (E) Reflected light intensities of the SPR system with and without a gold film.

A microscope coverslip with a gold film on one side was then pasted onto the hypotenuse surface of the prism by a drop of refractive index matching oil. A rubber O-ring glued on a piece of Plexiglas was pressed onto the gold film by a screw. The gold film, the Plexiglas and the rubber O-ring constructed a chamber with a volume of 250 μl. Water and biological samples could be injected into the chamber from a notch on the Plexiglas body. A yellow ruler in inches shown in the figure was a size reference. An AAA battery shown in figure 6.2(B) was another size reference for the prism, the O-ring chamber, the coverslip with a gold film and two solar cells.

Figure 6.2(C) is a photo of a vacuum ion sputter used to deposit gold films on coverslips. The sputter was a product of a company of the Chinese Academy of

Sciences in Beijing. Two square analogical meters on the front panel of the sputter were used to measure vacuum and sputtering current respectively. There was a needle valve between the two meters, and it was used to adjust the vacuum during the sputtering process. On the right of the front panel, there was a red power switch and a rotatable knob which was used to set the sputtering time. The sputtering chamber was a hollow glass cylinder sitting on the top of the equipment. A negative electrode (cathode) was in the black cover on the top of the cylinder. A gold foil was connected to the cathode. The diameter of the gold foil in the glass chamber was 10 cm. Four coverslips could be put on a positive electrode (anode) in the vacuum chamber.

After turning the power switch to the ON position, an oil vacuum pump began to pump air out of the chamber. In about 30 min, the pressure in the chamber went down to 4 Pa (Pascal), the sputtering process was then started by pushing a small red button on the front panel. By keeping the vacuum at 4 Pa and the sputtering current at 5 mA, a gold film (~ 50 nm) could be deposited on each of four coverslips in 200 s. This gold film deposition method is called cathodic sputtering [2] invented in 1852 by Sir William Grove. Ionized gas particles with positive charges in the chamber were accelerated by a high voltage to bombard the gold foil, so that tiny gold particles could be sputtered away from the gold foil. These gold particles fell on the coverslips and accumulated into thin gold films. We could also use the thermal evaporation method to do the same job. Gold films made by thermal evaporation method are very smooth and very shiny because the gold particles are smaller. Hundreds of coverslips could be deposited with high quality gold film in one batch. However, the equipment for thermal evaporation is very expensive. It takes about four hours to reach extreme low vacuum by three-stage vacuum pumps for the thermal evaporation chamber.

A red curve about the ZF3 glass refractive index and light wavelength is shown in figure 6.2(D). The data of the curve were obtained by measuring minimum deviations of monochromatic light waves on a prism spectrometer. Those monochromatic light waves were nine mercury spectral lines and a sodium spectral line (589.3 nm). Cauchy's equation of ZF3 is presented in the figure. The equation came from a nonlinear regression of these data. From this equation, the refractive index of ZF3 at 650 nm was found to be 1.7114. The refractive index of ZF3 can be found in a database published on https://refractiveindex.info/, and it is 1.7109.

How should the kind of glass to make a prism for an SPR system be chosen? Should we use a glass with a higher or a lower refractive index? According to the formula (3.34), the relationship between $\Delta\theta_{\mathrm{SPR}}$ and the refractive index change of a low density medium in an SPR system with Otto configuration can be estimated using the following formula

$$\Delta\theta_{\mathrm{SPR}} \approx \frac{\Delta n_1}{n_0 \cos(\theta_{\mathrm{SPR}})}\sqrt{\frac{\varepsilon'_{2r}}{n_1^2 + \varepsilon'_{2r}}} \tag{6.1}$$

From the data shown in table 3.1, we could see that a prism with lower refractive index (n_0) gave bigger values of θ_{SPR}, so the values of $\cos(\theta_{\mathrm{SPR}})$ were smaller.

Corresponding to the same value of Δn_1, an SPR system using a prism with a lower refractive index (n_0) gives a bigger $\Delta\theta_{SPR}$ value, so it is more sensitive. BK7 glass may be the best choice for prisms of SPR biosensors. It is easy to find coverslips made of BK7 glass for sensing chips with refractive index matching oil. But an SPR system with a low refractive index prism produces 'broad' SPR spectra at much bigger incident angles (>71°) in water. To precisely find θ_{SPR} values using software, we need sharper SPR spectra. Sometimes we have to use prisms made of glass with higher refractive indices.

There are two curves in figure 6.2(E). They were the relationships about reflected light intensity versus the incident angle of the homemade SPR system shown in figure 6.1. The blue curve is for the system without a gold film in the air, and the red curve is for the system with a gold film on a coverslip in the air. The SPR adsorption valley was not deep because the gold film was thicker than 50 nm. Measuring data for an SPR spectrum took more than half an hour. Some improvements were needed to make the SPR system more useful.

6.2 Modifying an x-ray diffraction goniometer for an automatic SPR biosensor

A second-hand x-ray diffraction goniometer was modified to make an automatic SPR biosensor, from which hundreds of data points for an SPR spectrum could be automatically acquired in about a minute. Electronic components, i.e. solid state relays, operational amplifiers, a 2-pole-3-throw switch, and two diodes were used to modify the goniometer so that it could be controlled by an IBM PC/XT personal computer with an AD/DA board.

6.2.1 The structure of the automatic SPR biosensor

The structure of the automatic SPR biosensor is shown in figure 6.3. An x-ray tube was replaced by a semiconductor laser diode, and an x-ray detector was replaced by a solar cell installed on a bigger disk with a rotation speed of 1.00° s^{-1} (figure 6.3(B)). A prism was set on a sample stage with a rotating speed of $\frac{d\varphi}{dt} = 0.50° \, s^{-1}$. Since the prism table and the bigger disk were coaxial, the solar cell could trace the reflected light automatically. Metal gears inside the goniometer guaranteed that the detector rotation speed was the double of the prism stage speed. This kind of equipment for tracing reflected light is a 'θ–2θ goniometer'.

The axis of the prism stage was virtually a vertical line on the hypotenuse surface of the prism. As shown in figure 6.3(B), the angle between the laser and the hypotenuse surface of the prism was φ. It was an outside incident angle to distinguish it from the 'inside incident angle' θ, the angle between the refracted laser ray and the normal of the prism hypotenuse surface. The two incident angles had an explicit relationship

$$\theta = 45° + \arcsin\left[\frac{1}{n_0}\sin(45° - \varphi)\right] \tag{6.2}$$

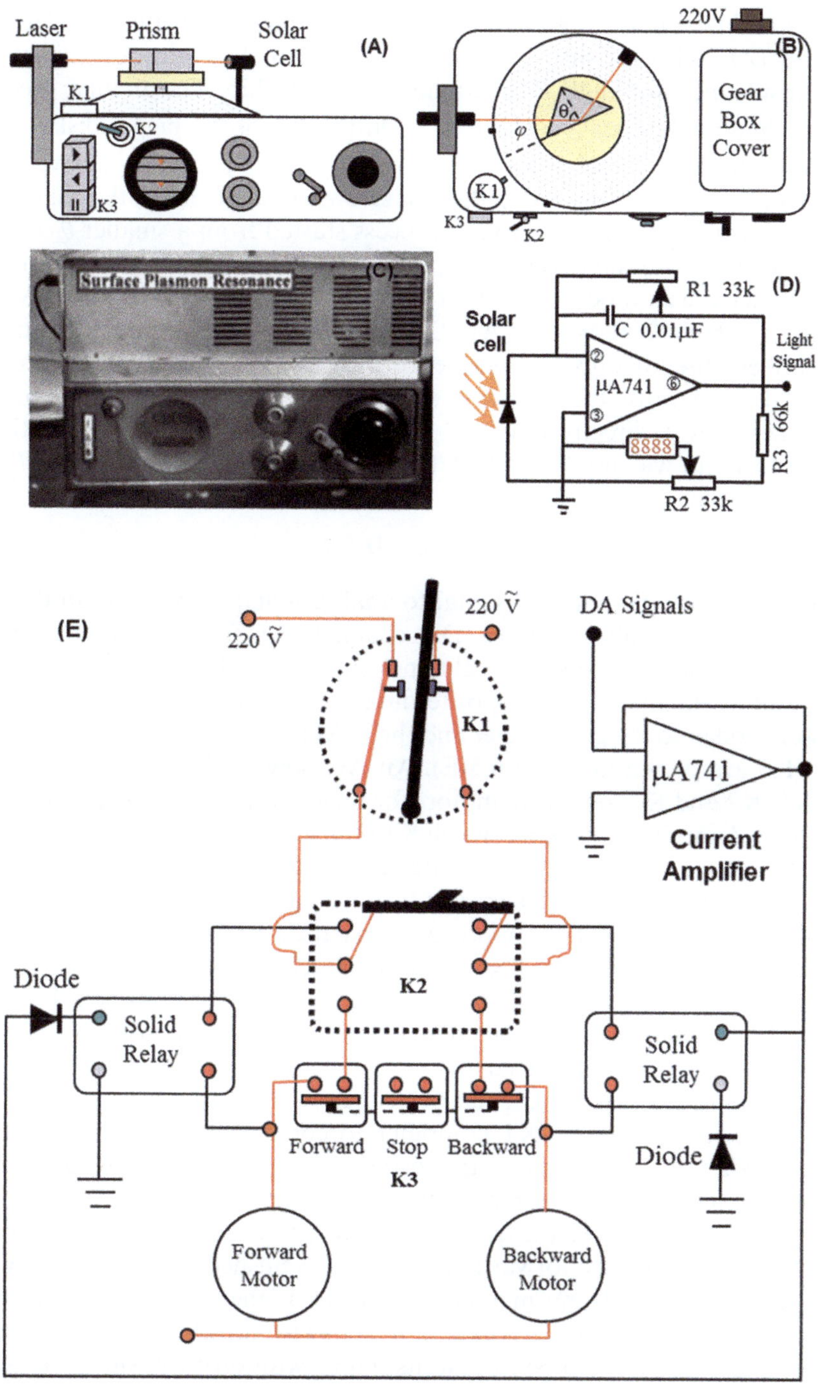

Figure 6.3. The structure of an automatic SPR biosensor based on a x-ray θ–2θ goniometer. (A) Front view. (B) Top view. (C) A photo of the SPR biosensor. (D) An operational amplifier μA741 connected to the solar cell for light detection. (E) An electric circuit for automatic control of the goniometer by a computer.

This formula could be replace by a simpler formula if φ is between 25° to 65°, and $n_0 = 1.795$ (ZF7 glass prism): $\theta = -0.552\varphi + 69.846$. The outside incident angle φ and the inside incident angle θ have a linear relationship. The angular speed of the inside incident angle was $\frac{d\theta}{dt} = 0.276° \text{ s}^{-1}$. On the rim of the bigger disk, there were two screws which were used to set an incident angle scan region. Whenever a lever in a switch K1 was pushed by any of the two screws (figure 6.3(B)), the rotation of the goniometer would be stopped. A scan process started from a smaller θ_0 or a bigger φ_0, and the clockwise direction is the positive turning direction of the prism table. The goniometer was modified so that it could by controlled by an IBM PC/XT computer in which there was an AD/DA board. Although we could read incident angle numbers directly on scales on the front panel of the instrument, the computer could not do so because there was not an angular sensor in the old system. Incident angle values were indirectly obtained by using the rotation speed and time. The following formula was used to calculate incident angle values in the AD/DA data acquisition system for SPR spectra

$$\varphi(t) = \varphi_0 - 0.50t \tag{6.3}$$

There was only one channel for digital to analog signal conversion on the AD/DA board. The output voltage of the DA channel was from −5 V to 5 V with a resolution of 12 bit. To use the DA channel output signal to control the goniometer rotation automatically in two opposite directions, two diodes and two solid state relays were added to the goniometer and their electric connections with switches K1, K2 and K3 are shown in figure 6.3(E). An AC power of 220 V was connected to switch K1, K2 and K3 to control motors for forward turning and backward turning respectively. A mechanic striking switch K1 on the top of the equipment was usually in the ON position for both of its two directions. At the beginning of an angular scan process, the right channel for backward turning was cut off, as shown in the figure; and at the end of the angular scan process, the left channel was broken by a screw on the rim pushing the lever to the opposite direction. A double-pole-three-throw switch K2 on the front panel of the goniometer had three positions. In its middle position, the goniometer was powered off. This was a manual position of the system. The goniometer could be rotated by hand through a crank for moving the two screws on the rim to set an angular scan region. The up position of K2 was an automatic position for a PC/XT computer to control the goniometer. The down position of K2 was a semi-automatic position. The switch K3 had three keys, and they were not independent. Pushing down any of the three keys would lock it in the ON position and un-release the other locked key. For example, if the K2 was in its semi-auto position and the forward key of K3 was pushed down, the backward key would jump back to its normal position (OFF), the forward motor was then connected to the AC power to rotate in a clockwise direction for a SPR spectrum scan. The goniometer could be reset to its initial position by pushing the backward key of K3.

Two solid state relays were used to bypass the switch K3 for automatic control of the goniometer by the computer. When the output of the DA channel was +5 V, an analog current flew into the left relay through the left diode. The left relay was ON to let the AC current drive the forward motor. But the right diode would block the analog current to flow into the right relay, so the right relay was off. When a backward rotation of the goniometer was needed, the DA channel sent out a −5 V signal, the left relay was OFF and the right relay was ON. However, the output current of the DA channel on the AD/DA board was not strong enough to drive these relays. A general-purpose operational amplifier μA741 was used to be a current amplifier. It was an 8 pin integrated circuit (IC) chip powered by ±12 V. Another μA741 was used to amplify the solar cell analog voltage shown in figure 6.3(D). A variable resistor R1 (33 kΩ) was used to adjust the amplification of the IC chip. By adjusting the resistance of R1, the maximum output of the amplifier could be set to 5 V. The AD channel of the AD/DA board could convert any analog voltage signal of −5 V to 5 V into a 12 bit digital signal with a number from 0 to 4095. A capacitor C (0.01 μF) was used as a noise filter. When the prism refractive index is higher than that of a coverslip with a gold film in a SPR sensing chip, light reflected from the two interfaces of the coverslip will interfere with each other to produce noises. If water was used in the chamber of the SPR system, the interference noise could be strong enough to distort the SPR spectra. By choosing the optimum capacitance of the capacitor, the noise could be filtered. A digital voltage meter was connected to the voltage amplifier for the convenience of setting the amplifier without turning on the computer. Its measurement region was 0–2 V. Two resistors R2 and R3 were used to set the digital voltage meter working properly.

On the AD/DA board, there was a six-bit DIP switch to set hexadecimal addresses of a memory in a region from 0200 to 03FC for data reading/writing. With the Borland Turbo BASIC programming language, the following short program instructed the AD/DA board to output a 5 V analogic signal. By changing FF to 00 and 0F to 00, the short BASIC program would send −5 V out of the DA channel. Any variable with a percentage symbol '%' at the end of its name is an integer. A number starting with '&H' is a hexadecimal number. Any sentence starting with a single quotation mark is a text region in BASIC programs explaining the program

```
'————————————————A Turbo BASIC Program for DA 5V————————————————
10 Address%=&H280             'set the address of the AD/DA board
20 Out Address%+3, &HFF       '255 or &HFF is written into (&H283)
30 Out Address%+4, &H0F       '15 or &H0F is written into (&H284)
40 End                   'They are combined as &H0FFF or 4095 for DA 5 V
'————————————————————The End of the Program————————————————————
```

There were 32 AD channels on the AD/DA board with a 12 bit resolution for analog signals from −5 V to 5 V. By writing an integer (0–31) into the memory at

(Address%), we could choose which AD channel would be used. After an AD conversion process was successfully done, an integer number 128 would be written to the memory at (Address% + 5). Line 30 and 40 in the following BASIC program were used by the computer to check if the AD conversion process was successfully done. If it was not, the program went back to line 30 to repeat the poll, otherwise read the AD conversion results in line 50 and line 60. The low 8 bit of the digital number of the analog signal was then stored in the memory at (Address% + 1), and the high 4 bit of that was stored in the memory at (Address% + 2). The following short BASIC program was for DA conversions. These short programs were integrated in a much longer BASIC program to record and display SPR spectra on screen. All data were written to text files stored on the hard disk of the computer.

```
'————————————————A Turbo BASIC Program for AD————————————————
10 Address%=&H280                    'the address of the AD/DA board
20 Out Address%, 0                   'Choosing DA channel 0
30 Poll%=Inp(Address%+5) AND 128     'Is the AD conversion ready?
40 If Poll%<>128 Then 30             'If not, go back to 30 to poll again
50 Low8Bits%=Inp(Address%+1)         'Input the low 8-bits in (&H281)
60 High4Bits%= Inp(Address%+2) AND &H0F     'Input the high 4-bits
70 AD%=256* High4Bits%+ Low8Bits%           'Digital data
80 Voltage=(AD%-2047)*10/4096               'The analog voltage
90 End
'————————————————————The End of the Program————————————————————
```

6.2.2 An application of the automatic SPR biosensor

An application of the automatic SPR biosensor was about interactions of rabbit C-reactive protein (rCRP) with phospholipid molecules. Antibody IgG is a popular protein sample in biosensor applications. It is a biological weapon in our adaptive immunological system. Although an adaptive immunological system in an animal is very specific and powerful, it takes time (3–4 weeks after inoculation) for the system to produce antibodies. Sometimes, the adaptive immunological systems are not quick enough to provide protection. There is an innate immune system in each living animal [3]. It quickly, and non-specifically, responds to any invasion. C-reactive protein (CRP) is a weapon in the innate immune system, and it is a major acute phase reactant in most mammalian species. In 1930, Tillett and Francis [4] found that sera from patients with acute inflammations reacted with C polysaccharide (CPS), a carbohydrate from capsules of pneumococci. The substance in patients' blood was found to be a protein in 1940 by Abernethy and his colleagues [5]. This kind of protein could tolerate temperatures up to 65 °C, and its reactions with CPS needed calcium irons (Ca^{2+}). This protein was purified and called 'C-reactive protein' in 1941 by MacLeod and Avery [6, 7]. By crystalizing the CPS–CRP complex in 1947, McCarty made it possible for us to see this mysterious protein under the microscope [8]. The 3D crystal structure of human CRP was published in 1996 by Shrive [9]. Now CRP gene sequences, protein sequences, and 3D crystal structures can be found in databases on the Internet. However, CPR's physiological functions are still not completely known.

CRP is an ancient protein [10, 11] existing in mammals, birds, reptiles, amphibians, and even in invertebrates such as horseshoe crabs. Usually a CRP molecule has five identical subunits [12], and its subunit's molecular weight is about 23 kDa. The molecular weights of human CRP and rabbit CRP are 135 kDa and 120 kDa respectively. The binding of a CRP subunit to a phosphocholine (PC) molecule is mediated by a calcium ion in a hydrophobic pocket of the subunit [13]. Residue phenylalanine 66 is in the center of the pocket, and residue glutamic acid 81 is in a position to interact with a choline group of the PC molecule. In a mammalian cell membrane, there are a lot of phospholipid molecules. Each phosphatidylcholine molecule has two hydrophobic lipid tails and a PC head. CRP molecules only bind with necrotic cells and invading microbes [14]. Using three synthesized lipids [15] with longer polar heads (figure 6.4) and the automatic SPR biosensor, information about steric hindrance of rCRP binding to phospholipid monolayers was investigated [16]. Between lipid tails and the PC head, a chain of 8, 6 and 3 atoms was inserted respectively, so these synthesized lipids were named as 'DS8PC', 'DS6PC' and 'DP3PC'.

Rabbit C-reactive protein was purified from the sera of rabbits by affinity chromatography on columns with phosphorylethanolamine–Sepharose 4B, a Sigma-Aldrich product. The rabbits were in acute phase inflammation when they were sacrificed. A photo of SDS-PAGE (sodium dodecyl sulfate–polyacrylamide gel

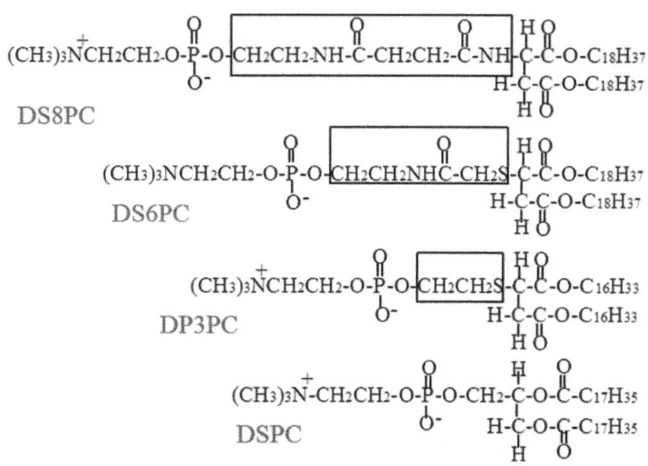

Figure 6.4. Chemical formulae of three synthesized lipids DS8PC, DS6PC, DP3PC and a natural phospholipid molecule distearoyl phosphatidylcholine (DSPC).

electrophoresis) and a photo of G-PAGE (gradient polyacrylamide gel electrophoresis) are shown in figure 6.5(A). They were used to measure the molecular weight of a subunit of rCRP and the molecular weight of a whole rCRP respectively. Molecular weights of the protein markers are listed for these photos. The marker proteins are: myosin 212 kDa, β-galactosidase 116.3 kDa, phosphorylase B 97.4 kDa, bovine albumin 66.2 kDa, ovalbumin 42.7 kDa, aldolase 40.0 kDa, carbonic anhydrase 30 kDa, and myoglobin 17.5 kDa. These gels were stained with Coomassie Blue. From the SPD-PAGE photo, the molecular weight of a subunit of rCRP was found to be 23 kDa. Because the pentamer rCRP molecule is a disk shape, not a sphere shape, the molecular weight of the rCRP measured by gradient polyacrylamide gel electrophoresis was a little higher.

A natural lipid DSPC and each of the three synthesized lipids were dissolved in an organic solvent (chloroform: methanol 3:1 volume ratio) respectively. On the water surface of a Langmuir–Blodgett trough with a water surface area of 45×9 cm^2, a mixture of DSPC and a synthesized lipid in 10 μl was dispersed, and then compressed to form a lipid monolayer to a surface pressure of 35 mN m^{-1} (millinewton per meter). The lipid monolayer on the water surface was then horizontally transferred to the surface of a coverslip with a gold film deposited. The coverslip with a gold film and a lipid film is a sensing chip for rCRP. After the

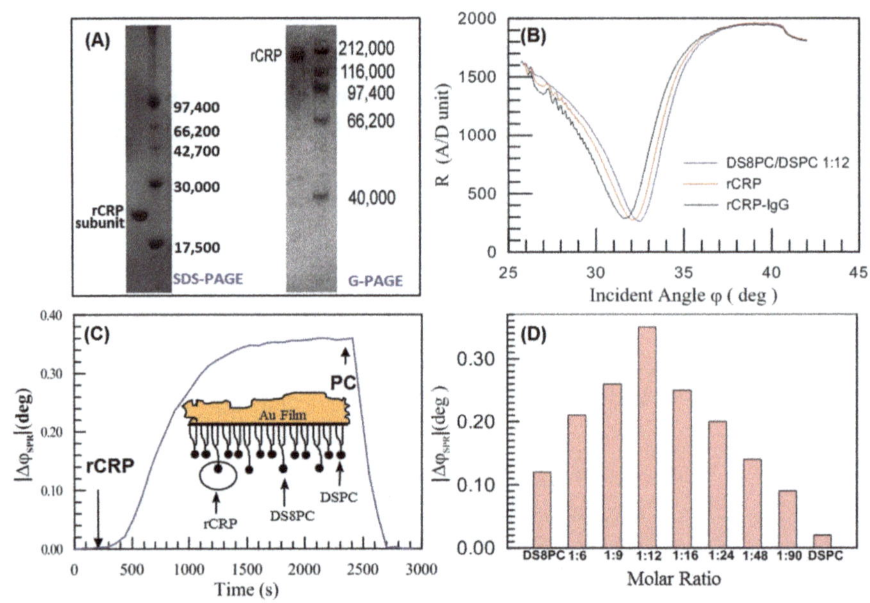

Figure 6.5. (A) photo SDS-PAGE about rCRP subunit molecular weight measurements and a photo of G-PAGE for rCRP molecular weight measurements. (B) SPR spectra of a sensing chip with DS8PC/DSPC (1:12), rCRP, and anti-rCRP IgG. (C) The specific binding rCRP to the sensing chip could be eluted by adding PC to the chamber. (D) The effects of the molar ratio of DS8PC/DSPC to $|\Delta\varphi_{SPR}|$.

sensing chip was installed in the SPR biosensor, the sample chamber was then filled with 150 µl buffer (Tris–HCl 20 mM, pH 7.4) containing 2 mM EDTA (ethylenediaminetetraacetic acid). Then rCRP (2 µl, 0.7 mg mL^{-1}) was injected into the chamber, no change of θ_{SPR} was observed. Three SPR spectra were presented in figure 6.5(B). Since one EDTA molecule binds with one calcium ion, there were free Ca^{2+} ions in the chamber after a 50 µl buffer containing 10 mM $CaCl_2$ was injected into the chamber. SPR spectra shifted (θ_{SPR} increased or φ_{SPR} decreased) only when DS8PC molecules were present on sensing chips with Ca^{2+} irons in the chamber. From figure 6.5(C) we could see that the binding of rCRP on a DS8PC/DSPC (1:12) monolayer saturated in 40 min. Then a 10 µl buffer containing 20 mM PC was added to the chamber, φ_{SPR} quickly shifted almost to the position before the $CaCl_2$ was added. If the rCRP liquid was first incubated with PC, then the rCRP could not react with sensing chips containing DS8PC. As the reaction of rCRP with DS8PC/DSPC monolayers was dependent on Ca^{2+}, and the reaction could be inhibited by PC, the interactions in the above experiments were specific. The black curve (left) in figure 6.5(B) is the SPR spectrum of a sheep anti-rCRP IgG film over the rCRP/lipid/Au system. Optical interference noises can be seen on the SPR spectra. A column graph in figure 6.5(D) showed the changes of φ_{SPR} or θ_{SPR} were dependent on the molar ratios of DS8PC to DSPC on sensing chips. The optimum molar ratio was found to be 1:12.

A molecule of DS8PC on a sensing chip acts as a receptor to a subunit of rCRP. It is reasonable to think that the denser the receptors, the more rCRP would bind to a sensing chip. The receptor density must have an optimum density, by which there was a maximum binding of the rCRP molecules on the sensing chip. But higher receptor density would not increase rCRP binding to the chip because of lateral hindrance coming from the geometrical sizes of rCRP protein.

At a molar ratio of 1:12, the DS8PC/DSPC monolayer had maximum specific binding to rCRP ($\Delta\theta_{SPR} = 0.193°$). A formula about protein surface concentration and SPR resonant angle change was obtained from numerical simulations: $\Delta\Gamma = 10\Delta\theta_{SPR}$. If a molar mass (Mol) of a sample molecule was given in a unit of g mol^{-1} or Dalton, then we could count how many protein molecules per square millimeter on a monolayer, and then find the average area [$Å^2$] of a protein molecule on the surface: $s = \frac{1}{60.23}\frac{\mathrm{Mol}}{\Delta\theta_{SPR}}$. With the molar mass of rCRP 1.2×10^5 g mol^{-1} it was easy to find that the average area occupied by a rCRP molecule over the lipid monolayer (DS8PC/DSPC 1:12) surface was 10 400 $Å^2$. From electron microscope observations [17], the diameter of a rCRP molecule was found to be about 100 Å. Therefore, the above SPR experimental results matched well with the diameter square of the rCRP molecule. This meant that the DS8PC/ DSPC monolayer with a molar ratio of 1:12 could be almost completely covered by rCRP molecules. In the above experiments, DP3PC/DSPC and DS6PC/DSPC monolayers could not react with rCRP. Flexible polar heads of DS8PC molecules in the DS8PC/DSPC monolayer might protrude deep enough to reach PC binding sites in rCRP molecules. From chemical formulae of DS8PC and DSPC, the depth of a PC binding pocket in a rCRP molecule was estimated to be about 10 Å.

6.3 A portable SPR biosensor

To design a SPR biosensor with a monochromatic light source, designers have to find way for a light detector to trace the reflected light. Using a θ–2θ goniometer in the above automatic SPR system is a solution. But even a second-hand x-ray diffraction goniometer is very expensive and not easy to find. Commercial SPR biosensors (BIAcore 1000) use focused light with different incident angles and a linear CCD array to detect reflected lights at different angles simultaneously. There is a third way to trace reflected light [18].

6.3.1 The structure of a portable SPR biosensor

Figure 6.3(B) is the top view of the automatic SPR biosensor discussed in the last section. In which the distance between the solar cell and the light reflection point in the prism was about $r = 200$ mm. If the incident angle change ($\Delta\varphi$) is 20°, the arc length the solar cell has to be moved is $2r\Delta\varphi = 140$ mm. By directly attaching a solar cell to a side of the prism, the r value can be reduced to 12.5 mm for a prism with two equal sides each with a length of 25 mm. In that case, the moving distance of the reflected light on the side of the prism is less than 9 mm. By using a solar cell with an area bigger than 10×10 mm^2, a trace system for the reflected light is not needed. A photo of a portable SPR biosensor is shown in figure 6.6, in which a solar cell with an area of L40 mm × W20 mm was directly attached to the holder of the prism. Reflected light would not move out of the solar cell. In the simplified SPR system, the reflected light intensity curves are a little different from real SPR spectra, but the resonance angles should be the same if the SPR spectra are sharper.

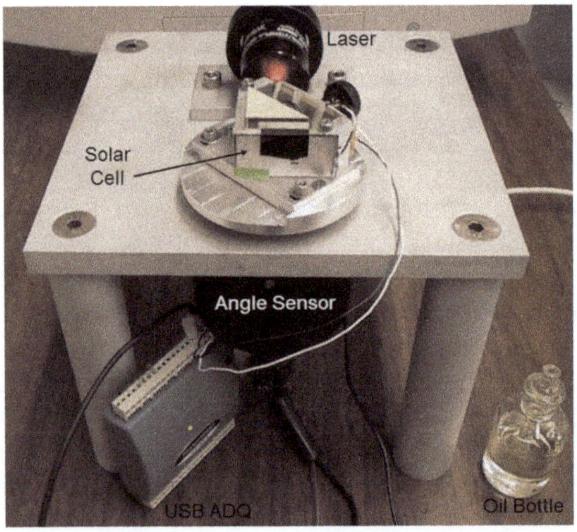

Figure 6.6. A photo of the portable SPR biosensor.

Since an angle sensor is added, the SPR biosensor can record incident angle and reflected light intensity simultaneously, so that a uniform rotation of the prism is not needed. The small disk holding the prism can be rotated manually to obtain SPR spectra. In the portable SPR biosensor shown in figure 6.6, a vertical shaft of an angular sensor was connected with the prism stage. The angular sensor (Model: 0603-0001) was a product of a Canadian company: Intertechnology Inc. It was an angular displacement transducer with precision differential capacitors without edge effects. When powered by ±15 V, it linearly converts its shaft angular displacement to a DC voltage signal with a resolution of 100 mV per degree in a region from −60° to 60°. With the angular sensor, we could measure the incident angle with a resolution of 0.01° if our DC voltage measurement resolution of is 1 mV.

The light source of the SPR system was a semiconductor laser, a product of Melles Griot (Model: 06DAL101). The wavelength of the laser was 650 nm, and its output power was 0.9 mW. The laser was a linear polarized light source with a polarization extinction ratio of 100:1. Its beam was rectangular. The laser was horizontally set on an aluminum table ($300 \times 300 \times 25$ mm^3) so that the rectangular shape of the laser spot was vertical for a p-polarized incident light. Four legs were used to support the aluminum table. Their heights were 200 mm.

In the photo, under the table beside the left-front leg, is a USB data acquisition device made by Measurement Computing (Model: USB-1608FS). Its eight AD channels could be independently configured for simultaneous data acquisitions (DAQ) with a resolution of 16 bits. Using software, their analogical signal input region could be set to: ±10 V, ±5 V, ±2 V or ±1 V. Channel 0 was used for the solar cell signal. Since the maximum voltage of the solar cell is 0.5 V, the input region of the channel was set to be from −1.0 V to 1.0 V with an accuracy of 0.68 mV. No extra amplifiers were used in the system. Channel 1 was used to measure the incident angle signal from the angular sensor whose output region was −6 V to 6 V. The input region of the channel was set to be −10 V to 10V with an accuracy of 5.66 mV. The angular resolution of the SPR system was 0.0566° (= 3.4′). To improve the angular resolution, we could set the input region of channel 1 to be −5 V to 5 V since the outside incident angles for SPR spectra in water were from 20° to 50°. The angular resolution of the system could be 0.0299° (1.8′).

A LabVIEW™ program interface and its source code are presented in figure 6.7. On the left of the interface, there was a waveform chart and a meter to show the reflected light intensity in mV and the incident angle φ in degrees. Both of these would be changed with time if the prism table was rotated manually after a button with a big arrow on the program interface was clicked. Each of them had a digital indicator to display exact values of analog signals from the two sensors. There was a digital control called 'Interval(s)' on the interface. Users could input data there to choose DAQ time intervals in units of seconds. In the same line on the right, there was a 'STOP' button. The DAQ process would be stopped if this button was clicked. Under the STOP button was an x–y plot which combined the incident angle information and the reflected light intensity information into a SPR spectrum.

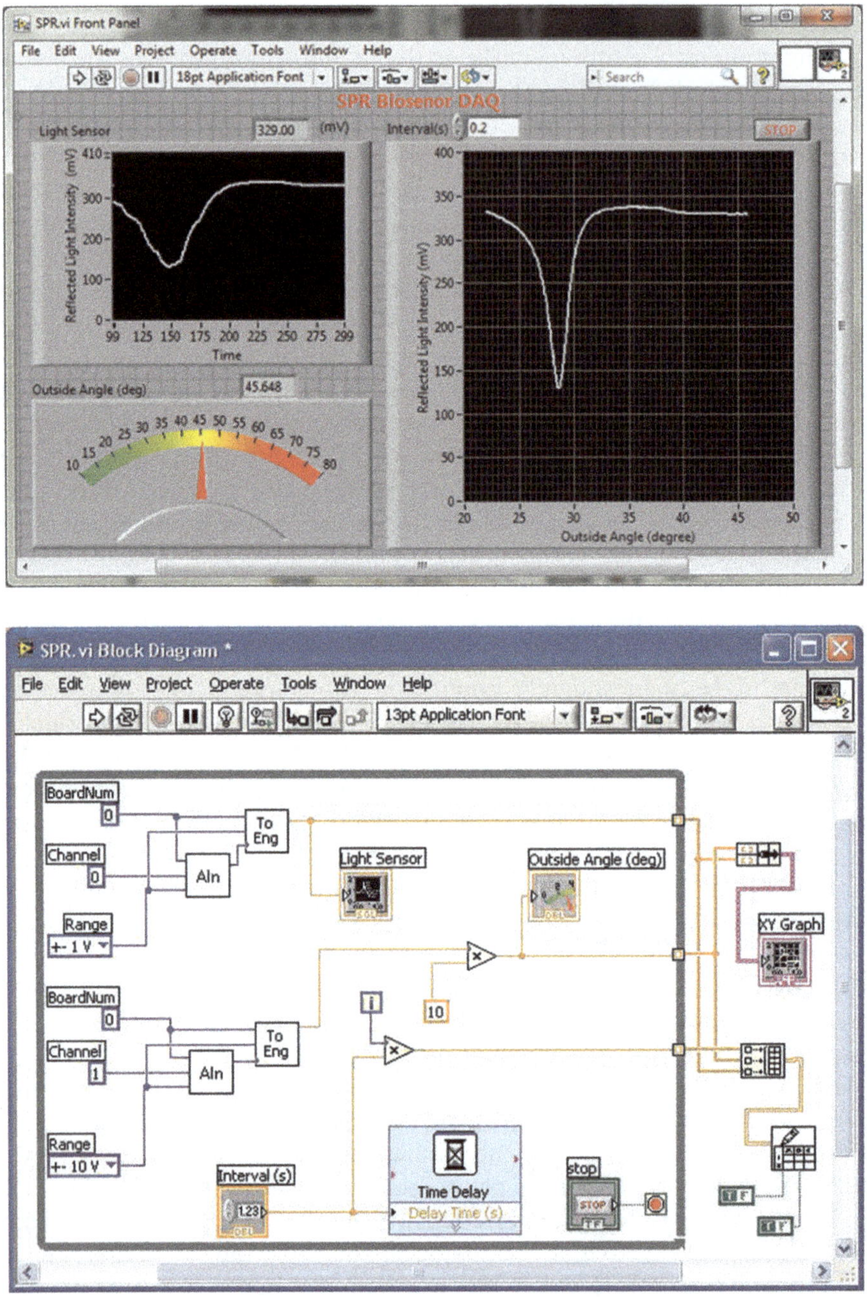

Figure 6.7. A Labview program interface and its source code for DAQ of the portable biosensor.

Since LabVIEW is a graphical programming language, its source code is also graphical with many icons connected by wires (figure 6.7). A rectangular area with a thick gray frame is a loop for the integer '*i*' inside the square. The loop repeats all icon jobs inside the frame with an interval of 0.2 s. All data shown in the waveform chart and the meter were stored in computer memory. Once the STOP button was clicked, the loop for DAQ would stop. All data were combined to draw a SPR spectrum, and these data would then be written to a data file with three columns: time (t), outside incident angle (φ) and reflected light intensity (R). Usually scanning a SPR spectrum takes about 80 s, so that a data file for a SPR spectrum has about 400 rows.

6.3.2 An application of the portable SPR biosensor

C-reactive protein is a biochemical weapon in our innate immune systems to fight infections by specifically binding to polysaccharides on bacterial cell walls. Plants also make this kind of protein weapon to prevent them from being eaten by animals and to fight fungal infections. This protein in plants is lectin. A protein molecule conjugated with sugar molecules is a glycoprotein, and a lipid molecule chemically linked with sugar molecules is a glycolipid. Sugar is a product of plants' photosynthesis. There are many kinds of sugar in nature. Single sugar molecules could polymerize to be fibers (wood), to be starch (food) to feed us, to be a backbone of DNA to carry our genetic information. Glycoprotein and glycolipids present their sugar chains or branches in our cell surfaces as our cells' identities to our immune systems. These short sugar chains are 'oligosaccharides'. Specific oligosaccharides presenting on the surfaces of our red blood cells decide our ABO blood types. Lectins from plants and bacteria could specifically recognize and bind with these oligosaccharides to agglutinate our blood. Bacteria make their own lectins to attach to their host cells in our bodies. Lectins reacting with specific terminal sugar residues might help us to find a way to cure cancers.

The portable SPR biosensor was used to investigate [19] Concanavalin A (ConA), a lectin specifically binding with mannose moieties or residues. ConA is a useful lectin extracted from jack beans. Each ConA molecule has four similar subunits. With a molecular weight of 26.5 kDa, a heavily glycated ConA subunit has 235 amino acid residues. Each of its subunits can bind with a mannose and/or glucose residue on an end of a oligosaccharide with the assistance of a calcium iron (Ca^{2+}) and a manganese iron (Mn^{2+}). The SPR investigation confirmed that ConA could facilitate *E. coli* binding to monolayers with mannoses.

A functional synthesized lipid molecular formula ($C_{25}H_{50}O_{10}S$) is shown in figure 6.8(A). Its hydrophilic end is a mannose sugar residue, and its hydrophobic chain end is a thiol group. The lipid molecular weight is 542 Dalton. The dashed blue curve in figure 6.8(A) is a SPR spectrum of a virgin gold film installed in the portable SPR biosensor with 100 µl ethanol in its sample chamber. Its SPR resonant angle θ_{SPR} is about 55.18°. After the synthesized lipid (dissolved in 100 µl ethanol to 4 mg mL^{-1}) was injected into the chamber, the thiol groups of lipid molecules interacted with gold atoms on the gold film to form a SAM over the gold film. The assembly process was monitored by recording reflected light intensity changed with time at θ_{SPR}. After 1.4 h, the chamber was washed three times and refilled with ethanol. The second SPR spectrum (thick black curve) was recorded and its θ_{SPR} was

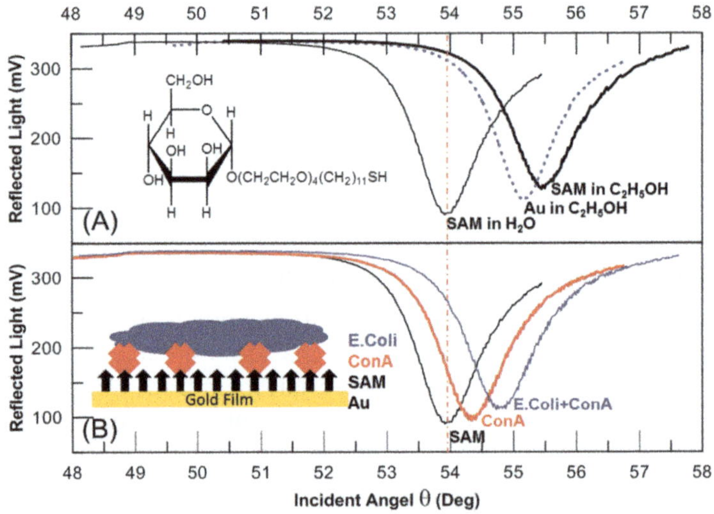

Figure 6.8. (A) SPR spectra for a synthesized lipid film (SAM) with mannose residues as receptors for ConA. (B) SPR spectra about ConA and *E. coli* k12. The noises on the spectra came from interferences of lights reflected from two surfaces of a coverslip of a sensing chip.

moved to 55.45°. The shift of the SPR resonant angle caused by the SAM in ethanol was 0.27°. The chamber was then washed and refilled with phosphate-buffered saline (PBS, pH 7.2, with 1.0 mM Ca^{2+} and 1.0 mM Mn^{2+}) for biological samples. The thin black curve on the left of the dashed curve was a SPR spectrum for the lipid film in water. Its SPR resonance angle was about 53.95°. The refractive indices of ethanol and water are about 1.36 and 1.33 respectively. Their difference is 0.03, and to the same lipid film, the $\Delta\theta_{SPR}$ is 1.50°. The SPR biosensor could be used to measure the liquid refractive index with a resolution of 0.0002 if its angle resolution was 0.01°.

The functional sensing chip with mannose residues as receptors could specifically react with ConA, but not with *E. coli*. Figure 6.8(B) shows three SPR spectra in PBS. The thin black curve was the same SPR spectrum shown in figure 6.8(A). The thick red spectrum in the middle was for the ConA film over the SAM after 40 min inoculation, and then the chamber was washed/refilled with PBS. Before washing, the final concentration of ConA in the chamber was 1.25 μM. The specification of the ConA to the SAM was verified by BSA and another lectin, *Erythrina cristagalli.* The SPR resonant angle of the ConA film over the SAM was 54.38°. Then if bacteria *E. coli* k12 (10^8 cfu mL^{-1}) were injected into the chamber, the SPR spectrum would not move. The bacteria could not directly bind to the ConA film over the SAM. But if a mixture of the bacteria and ConA were injected into the chamber, we could see the SPR spectrum moved to $\theta_{SPR} = 54.80°$ after 1.5 h incubation. In the chamber, the final ConA and bacteria concentrations were 1.25 μM and 10^8 cfu mL^{-1} respectively (cfu/mL: colony-forming unit per milliliter). Washing the chamber with PBS would not wash away the bacteria on the sensing chip. It is a mystery [20, 21] how and why ConA can facilitate *E. coli* binding to the ConA film over the SAM with mannose receptors on a gold film.

References

[1] Liedberg B, Nylander C and Lundström I 1983 Surface plasmon resonance for gas detection and biosensing *Sens. Actuators A Phys.* **4** 299–304

[2] Feigelson R S 2015 Crystal growth through the ages: a historical perspective *Handbook of Crystal Growth* ed T Nishinaga 2nd edn (Boston: Elsevier) ch 1 1–83

[3] Charles J and Janeway A 1993 How the immune system recognizes invaders *Sci. Am.* **9** 72–9

[4] Tillett W S and Thomas Francis J 1930 Serological reaction reactions in pneumonia with a non-protein somatic fraction of pneumococcus *J. Exp. Med.* **52** 561–71

[5] Abernethy T J and Avery O T 1941 The occurrence during acute infections of a protein not normally present in the blood: I. Distribution of the reactive protein in patients' sera and the effect of calcium on flocculation reaction with C polysaccharide of pneumococcus *J. Exp. Med.* **73** 173–82

[6] MacLeod C M and Avery O T 1941 The occurrence during acute infections of a protein not normally present in the blood: II. Isolation and properties of the reactive protein *J. Exp. Med.* **73** 183–90

[7] MacLeod C M and Avery O T 1941 The occurrence during acute infections of a protein not normally present in the blood: III. Immunological properties of the C-reactive protein and its differentiation from normal blood protein *J. Exp. Med.* **73** 191–200

[8] McCarty M 1947 The occurrence during acute infections of a protein not normally present in the blood: IV. Crystalization of the C-reactive protein *J. Exp. Med.* **85** 491–8

[9] Shrive A K, Gheetham G M T, Holden D, Myles D A A, Turnell W G, Volanakis J E, Pepys M B, Bloomer A C and Greenhough T J 1996 Three dimensional structure of human C-reactive protein *Nat. Struct. Biol.* **3** 346–54

[10] Pepys M B, Dash A C, Fletcher T C, Richardson N, Munn E A and Feinstein A 1978 Analogues in other mammals and in fish of human plasma proteins, C-reactive protein and amyloid P component *Nat. Struct. Biol.* **273** 168–70

[11] Hokama Y and Nakamura R M 1987 C-reactive protein current status and future perspectives *J. Clin. Lab. Anal.* **1** 15–27

[12] Osmand A P, Friedenson B, Gewurz H, Painter R H, Hofmann T and Shelton E 1977 Charactorization of C-reactive protein and the complement subcomponent C1t as homologous proteins displaying cyclic pentameric symmetry (pentraxins) *Proc. Natl. Acad. Sci. USA* **74** 739–43

[13] Thompson D, Pepys M B and Wood S P 1999 The physiological structure of human C-reactive protein and its complex with phosphocholine *Structure* **7** 169–77

[14] Volanakis J E and Wirtz K W A 1979 Interaction of C-reactive protein with artificial phosphatidylcholine bilayers *Nature* **218** 155–7

[15] Zhou Q, Gao Q, Tong J, Zhou Y, Mao J and Sui S-F 1998 Design, synthesis and characterization of artificial phospholipids as model membrane receptors for specific binding of rabbit C-reactive protein *Supramol. Sci.* **5** 795–802

[16] Sui S-F, Xiao C and Wang S-X 1999 Interaction of rabbit C-reactive protein with phospholipid monolayers at air:water interface *Colloids Surf. B: Biointerfaces* **15** 297–302

[17] Wang H-W and Sui S-F 1999 Pentameric two-dimensional crystallization of rabbit C-reactive protein on lipid monolayers *J. Struct. Biol.* **127** 283–6

[18] Xiao C and Zeng X 2010 Portable surface plasmon resonance biosensor *Google Patents* (Rochester, MI: Oakland University) p 31

[19] Shen Z, Huang M, Xiao C, Zhang Y, Zeng X and Wang P G 2007 Nonlabeled quartz crystal microbalance biosensor for bacterial detection using carbohydrate and lectin recognitions *Anal. Chem.* **79** 2312–9

[20] Gamella M, Campuzano S, Parrado C, Reviejo A J and Pingarrón J M 2009 Microorganisms recognition and quantification by lectin adsorptive affinity impedance *Talanta* **78** 1303–9

[21] Wang B and Anzai J-I 2015 Recent progress in lectin-based biosensors *Materials* **8** 5478

Lightning Source UK Ltd.
Milton Keynes UK
UKHW050853270619
345142UK00004B/62/P

9 780750 323826